TRIBOSYSTEM ANALYSIS

A Practical Approach to the Diagnosis of Wear Problems

TRIBOSYSTEM ANALYSIS

A Practical Approach to the Diagnosis of Wear Problems

Peter J. Blau

CRC Press
Taylor & Francis Group
Boca Raton London New York

CRC Press is an imprint of the
Taylor & Francis Group, an **informa** business

CRC Press
Taylor & Francis Group
6000 Broken Sound Parkway NW, Suite 300
Boca Raton, FL 33487-2742

First issued in paperback 2019

ISBN-13: 978-1-4987-0050-4 (hbk)
ISBN-13: 978-0-367-87137-6 (pbk)

Library of Congress Cataloging-in-Publication Data

Names: Blau, P. J., author.
Title: Tribosystem analysis : a practical approach to the diagnosis of wear problems / author, Peter J. Blau.
Description: Boca Raton : Taylor & Francis, 2016. | Includes bibliographical references and index.
Identifiers: LCCN 2015046369 | ISBN 9781498700504 (alk. paper)
Subjects: LCSH: Tribology.
Classification: LCC TJ1075 .B556 2016 | DDC 621.8/9--dc23
LC record available at http://lccn.loc.gov/2015046369

Visit the Taylor & Francis Web site at
http://www.taylorandfrancis.com

and the CRC Press Web site at
http://www.crcpress.com

I would like to dedicate this book to my wife, Evelyn,
and to the memory of my parents Jack Z. and Bella R. Blau
whose support and encouragement are greatly missed.

Contents

Preface xi

Author xv

1   What Is a Tribosystem? 1
    References 5

2   How Wear Problems Reveal Themselves 7
    2.1   Direct and Indirect Indications of Wear and Friction Problems 9
    2.2   Visual Inspection and Surface Imaging 10
    2.3   Vibration Monitoring and Acoustic Emissions 12
    2.4   Motor Current Signature Analysis 15
    2.5   Oil Analysis 15
            2.5.1   Sampling 16
            2.5.2   Chemical Analysis of Lubricants and Working Fluids 17
            2.5.3   Debris Particle Characterization 18
            2.5.4   Filtration, Magnetic Traps, and Ferrography 22
    2.6   Radionuclide Wear Detection 27
    2.7   Summary 27
    References 29

3   Types of Surface Damage and Wear 31
    3.1   Types of Surface Damage 33
    3.2   Types and Characteristics of Wear 37
            3.2.1   Wear Categories, Processes, and Mechanisms 40
            3.2.2   A System for Use in Distinguishing the Common
                    Forms of Wear 40
                    3.2.2.1   Sliding Contact (Category Code: S) 41
                    3.2.2.2   Repetitive Impact (Category Code: RI) 61
                    3.2.2.3   Contact Fatigue (Category Code: CF) 69
                    3.2.2.4   Wear with Chemical Attacks 78
    3.3   Material Dependence of Wear Patterns and Surface Appearance 80
    3.4   Wear Transitions 83
    3.5   Wear Diagnosis by Multiple Attributes in Context 85

Appendix 3A: Wear Nomenclature and Nomenclature Families 87
Appendix 3B: Resources for the Analysis of Wear Problems........................98
References 99

4 Tools for Imaging and Characterizing Worn Surfaces.......................103
    4.1   Light Optical Microscopy..106
           4.1.1   Specimen Mounting 108
           4.1.2   Specimen Cleaning 110
           4.1.3   Cross-Sectioning...112
           4.1.4   Taper-Sectioning...113
           4.1.5   Transmitted versus Reflected Light Illumination 114
    4.2   Stylus Profiling and Noncontact Topographic Imaging.................116
           4.2.1   Comparison of Roughness Data from Different
                    Measurement Methods...120
           4.2.2   Confocal Microscopy...121
    4.3   Scanning and Transmission Electron Microscopy 123
    4.4   Nontraditional Imaging Methods...124
           4.4.1   Scanning Acoustic Microscopy 124
           4.4.2   Thermography and Thermal Wave Imaging 127
    References 128

5 The Tribosystem Analysis Form 129
    5.1   An Overview of the TSA Form..130
    5.2   A Further Explanation of the Entries in the TSA Form 132
           5.2.1   A Detailed Description of the Components, Geometry,
                    and Materials 137
           5.2.2   A Description of the Operating Conditions 138
           5.2.3   A Problem Description...144
    5.3   Missing Information...148
    5.4   Tailoring the TSA to Practical Problems..150
           5.4.1   Piston Ring Groove Wear 150
           5.4.2   Vanes for Diesel Engine Turbochargers............................150
           5.4.3   Diesel Engine Valve/Valve Seat Wear...............................151
           5.4.4   Lightweight Rotors for Truck Brakes 151
           5.4.5   Gerotor Cases for Fluid Pumps 152
           5.4.6   Fuel Injector Plungers for Low-Sulfur Fuels....................152
           5.4.7   Wind Turbine Gearbox Bearings 153
           5.4.8   Human Teeth 153
    References 155

6 Wear Problem Solving—The Next Steps 157
    6.1   Options for Wear Problem Solving...158
    6.2   The Purpose for Tribotesting and the Type of Information Such
           Tests Can Provide..160

6.3 TSA Matching—Field versus Laboratory 165

    6.3.1 Case Study 1: Friction Coefficient versus Field
    Performance for Brake Lining Materials 168

    6.3.2 Case Study 2: Spectrum Loading Tribotests to Simulate
    Engine Bearings 171

    6.3.3 Variable-Condition versus Constant-Condition Tribotests 173

6.4 Wear Testing to Meet Specifications—Some Potential Pitfalls 173

References 175

Index **177**

Preface

There's an old joke about an old man who lost his car keys in a parking lot one night and decided to search for them in the center of the parking lot because the light was better there, not in the darker area near the corner where he actually thought he lost them. Several years ago, I was involved in a project to select an auger-type feed screw for a wood-chip-fueled power plant. The engineers were worried about abrasive wear as the auger screw transported wet wood chips upward and into the burner chamber. A laboratory-scale rotary stirring device was built and used to screen candidate weld overlay materials to recommend the best one. After construction of the plant, no auger wear problems were reported, but corrosion of the piping elsewhere in the plant caused the entire project to be halted. By focusing on one part of the problem, a potentially greater source of concern elsewhere in the system was ignored. This experience also reminds me of a limerick I wrote some years ago when the Soviet Union (USSR) still existed:

A man from the USSR,
Put ZDDP* in his car,
The engine ran great
For the good of the State,
But the tires wore out in an hour.

The point is that most scientists and engineers will, at some time in their careers, be faced with a friction or wear problem. Such multidisciplinary problems can be extremely challenging because our experience and training sometimes limits our perspective. Unfortunately, few receive formal training in tribology, the science and technology of friction, lubrication, and wear. Therefore, with a few notable exceptions, even technically trained persons in traditional fields are painfully ill-equipped to deal with the practical consequences of friction and wear and how they can affect the design, operation, and repair of mechanical devices. Few engineers have been formally trained in the selection of materials and lubricants, methods of tribotest selection, or the analysis and diagnosis of friction and wear problems. This

---

* ZDDP is a common antiwear additive in engine oils.

book was prompted by a need to provide a systematic approach to wear diagnosis that uses a tool which I call *tribosystem analysis* (TSA). Like other engineering tools, TSA has a place in the practicing engineer's toolbox, and its proper use can be helpful in seeing the "big picture" needed to solve wear problems.

The famous statesman Bernard Baruch supposedly said: "If all you have is a hammer, everything looks like a nail." Wear diagnosis and the tools used for wear testing are in some ways like this. Some laboratories may have purchased or built their own wear testing machines, and if a wear problem presents itself, they use the available apparatus to screen alternative materials or lubricants. The problem is that the type of wear that occurs may not be well simulated by the testing equipment or technical expertise that an organization has on-site. If the engineers recognize this shortcoming, they may seek help from an external testing laboratory, but the same concern still applies. While a given commercial wear testing laboratory may have a wide collection of equipment, they may still lack the best match of capabilities to solve the specific wear problem at hand. Performing a tribosystem analysis will not only allow wear problems to be defined systematically but will also help those tasked with their solution decide what test methods are appropriate.

Constructing a TSA is similar to, but not equivalent to, performing a root cause analysis. Rather, its primary function is to define the wear or friction problem in the context of its surrounding mechanical, chemical, and thermal environment. Once the problem has been defined by TSA, the solution can be approached by imposing changes in operating conditions, materials, lubricants, or even changing the machinery design. Consequently, this book takes a systematic approach to wear analysis and problem definition. It includes examples to illustrate some of the pitfalls and complexities which are not normally recognized by engineers who are presented with a tribology problem yet who may be unfamiliar with wear science and engineering.

Chapter 1 explains the meaning of the term *tribosystem*. This term is not unique to this book. In fact, it is defined in the ASTM International terminology standard for wear and erosion (ASTM G40). Unfortunately, the term tribosystem is not without a measure of ambiguity because it requires us to set certain boundaries. Broadly speaking, an entire automotive engine can be considered a tribosystem, but so can a piston ring and cylinder liner combination or a large-end connecting rod bearing deep within the churning heart of an engine. Therefore, some attention is given to establishing the boundaries of tribosystems for the purpose of the wear problem definition. This also leads to the concept of open and closed tribosystems.

Chapter 2 describes how wear problems manifest themselves. In some cases, a wear problem is obvious by visual inspection, but in others, wear is detected more subtly and indirectly, like noting discoloration of a lubricant sample, a changing trend in oil chemistry, rattling noises from a bearing, temperature rise, or an increased leakage of a seal. Wear and friction transitions, such as the onset of scuffing or galling, can occur suddenly and with little warning. Therefore, the ability to detect potentially damaging changes in interfacial conditions before it is too late is important for the operation of safety-critical tribosystems.

Chapter 3 presents a formalized approach for categorizing the various types of wear. In comparison with other approaches that are more mechanistically based, the current approach is based on the type of relative motion between interacting bodies, supplemented by observations. In defining the forms of wear and surface damage, attention is paid to historical terminology and therefore some ambiguous terms such as *scoring* and *scuffing* must be acknowledged. Using consistent terminology to identify the forms of wear in a tribosystem analysis is key to finding solutions based on successfully treated similar problems, selecting test methods, and adopting consistent approaches to wear problem solving.

Chapter 4 presents a "shirt-sleeve" overview of common tools for observing and characterizing wear and surface damage. In some instances, a simple hand lens or the naked eye can be sufficient, as opposed to employing an expensive, high-resolution electron microscope. The discussion stresses the importance of choosing the right tool for the right job, but also presents a variety of specialized techniques and hints, such as taper-sectioning and dynamic gray-level imaging. The use of oil analysis and trending, vibrational analysis, motor current signal analysis, ferrography, and debris morphology studies are also mentioned, along with references to provide additional depth.

Chapter 5 describes the parts of the TSA form and provides some examples of its use. While a complete description of a tribosystem may be impractical or even unnecessary, discovering what is not known about the tribosystem can be as useful as stating what is known. Indicating the type of wear is part of the TSA, and using consistent terminology, as described in earlier chapters, helps to provide a solid basis for wear test method selection. An essential element of the TSA is the history of the problem and a knowledge of what has or has not worked before in similar cases.

Chapter 6 includes some examples from my experience in addressing what options are available to solve wear problems. It provides information on how to apply the TSA concept to help select wear (and friction) test methods. Matching the TSA of a candidate test method to the tribosystem analysis of the problem components can enable better, more relevant problem solving, including the selection of materials and lubricants. Tribosystem analysis is a means to an end and not an end in and of itself. Like any tool, it has limitations, but its benefits can be powerful if used correctly.

Peter J. Blau

Author

Peter J. Blau, PhD, is currently the coeditor-in-chief of *Wear* journal and a consultant in tribology. He earned BS and MS degrees in metallurgy and materials science from Lehigh University (Bethlehem, Pennsylvania) and a PhD in metallurgical engineering from The Ohio State University (Columbus). His professional career began as a coinvestigator under the National Aeronautics and Space Administration (NASA) Apollo Lunar Sample Examination Program. Since then, he has spent nearly 40 years in basic and applied research and has held positions at the U.S. Air Force Materials Laboratory, the National Bureau of Standards (now the National Institute of Standards and Technology [NIST]), the U.S. Office of Naval Research, and Oak Ridge National Laboratory. A long-standing contributor to the International Conferences on Wear of Materials, he is a past conference chair and editor of several proceedings volumes. In 2011, the Wear of Materials Steering Committee established the Peter J. Blau Best Poster Award to be presented at each conference. Dr. Blau's work has earned his election as a Fellow of three societies: ASM International, ASTM International, and the Society of Tribologists and Lubrication Engineers. He also coholds several patents and R&D 100 Awards.

In addition to this book, Dr. Blau has authored or coauthored several books on microindentation hardness, friction, and wear. In the early 1990s, he worked with ASM International to develop the first *ASM Metals Handbook* volume devoted to friction, wear, and lubrication. Since 1979, he has participated in ASTM Committee G02 on Wear and Erosion, serving two terms as its chairman and also as chairman of Subcommittee G02.91 on Terminology. In addition, he has led and participated in the development of four standard test methods. Dr. Blau served as leader of the Tribology Research User Center at Oak Ridge National Laboratory and as a principal investigator for over 26 years. Since leaving Oak Ridge National Laboratory in 2013, he has been active in technical editing, writing, and consulting.

Chapter 1

What Is a Tribosystem?

Sooner or later, everyone experiences a friction or wear problem. What people choose to do about it is a different matter. You could simply ignore the problem or discard the worn-out device. If that is not an option, then you can replace the part or the entire device with a new one. As an engineer who needs to dig deeper into the problem, you may decide to improve the component design, try a different lubricant, or consider using alternative materials, surface treatments, finishes, or coatings. This book is intended to help those who need to do something about a wear (or friction) problem, whether it involves understanding how it happened, diagnosing the kind of wear that is occurring, developing or using a test to select alternate materials or lubricants, or completing a root cause analysis that could eventually lead to longer product life, improved design, and greater component reliability.

Wear problems belong to the multidisciplinary field of science and engineering that is called "tribology." Its name derives from the Greek word *tribos*, which in English means "I rub." The word *tribosystem* refers to a physical arrangement of two or more interacting structural parts, including the materials of which they are composed and the environment in which friction and wear is occurring.

ASTM terminology standard G40-15 [1] formally defines a tribosystem as follows:

> *tribosystem*, n.—any system that contains one or more triboelements, including all mechanical, chemical, and environmental factors relevant to tribological behavior.

And it goes on to define *triboelement* as follows:

> *triboelement*, n.—one of two or more solid bodies comprising a sliding, rolling, or abrasive contact, or a body subjected to impingement

1

or cavitation. (Each triboelement contains one or more tribosurfaces.) Discussion—Contacting triboelements may be in direct contact or may be separated by an intervening lubricant, oxide, or other film that affects tribological interaction between them.

Examples of triboelements include the following:

A ball within a rolling element bearing	A bearing race in a rolling element bearing
The separator in a rolling element bearing	A sanding disk on a woodworking machine
The shaft inside a journal bearing	A gear within an automatic transmission
A ceramic face seal in a water pump	A curling stone sliding on ice
A dental root canal file	The surface of a femoral head in an implant
A blade in an airport luggage carousel	A pipe carrying coal slurry
The sole or heel of a walking shoe	A slitter blade in a paper factory
Shearing rock layers in a moving fault	The hind legs of a "chirping" cricket

From a macroscopic point of view, a complex mechanical device like an engine, a pump, or a gearbox can be composed of more than one (sub)tribosystem, each of which contains a number of discrete triboelements. Thus, the definition of a tribosystem's boundaries is more obvious in some cases and less so in others. However, a working definition should include all the components and materials that surround the tribological interface of interest and that could influence its friction or wear behavior. That is, to establish the boundaries of a tribosystem, it is usually best to begin with the contacting surface(s) of interest and work outward.

In his book on wear analysis, Bayer [2] distinguishes between a macrotribosystem and a microtribosystem, with the latter being the contact zone experiencing friction or wear and the former containing the surrounding parts of the device. The exercise of defining a tribosystem's boundaries can be helpful when troubleshooting because it forces one to consider both internal and external influences.

One example of a complex tribosystem whose subassemblies can be considered tribosystems in and of themselves is the automatic transmission in an automobile. A conventional automatic transmission consists of a torque converter, a planetary gearset, clutches and bands, and the fluid system, which includes a pump and the associated valve body. Any assembly of individual gears (triboelements) could be considered a tribosystem, as could a clutch, or a seal, or a set of bearings that operate within that transmission. If a particular gear might be more problematic than others, then it would be appropriate to narrow the tribosystem of interest to include that gear and the surrounding gear or gears that come into contact with it. In bench-scale laboratory testing, the definition of tribosystem is more straightforward: The tribosystem is the test apparatus, the materials and any lubricants within it, and the immediate surroundings.

Figure 1.1 shows how a large and complex machine can be partitioned into a series of smaller tribosystems, each having its own functional requirements, surroundings, operating conditions, and materials. In this case, level 1 (not shown in the figure) would be the entire vehicle. Level 2 (one of several subtribosystems) shows just the engine. Owing to the large number of friction and wear interfaces involved, multiple types of wear can occur even within the same subtribosystem. For example, a piston ring in a ring groove in level 4 could experience microwelding, impact, rocking, circumferential slip, and fretting. Therefore, the process of conducting a tribosystem analysis is not limited to discovering only one dominant type of wear or surface damage, but rather identifying any and all relevant influences on the desired function of that tribosystem.

A similar approach to that in Figure 1.1 has been taken to categorize test methods that simulate practical tribology problems. For example, German standard 50

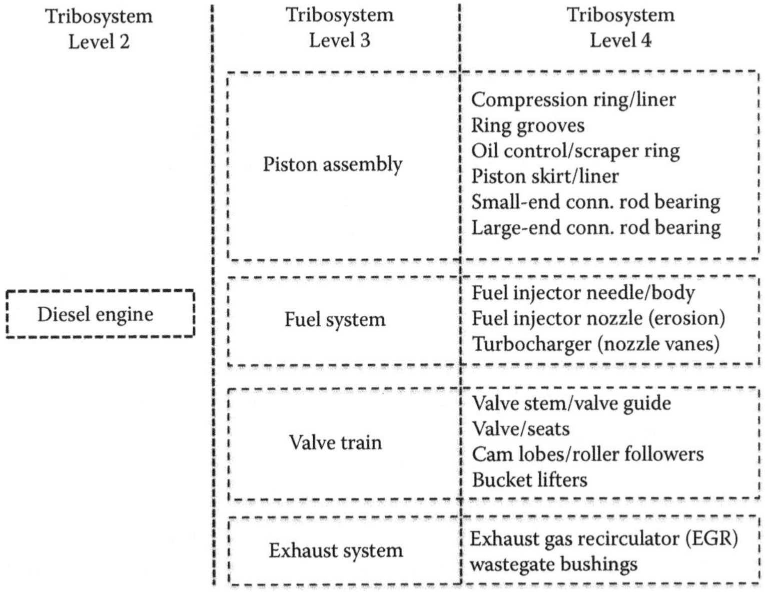

Tribosystem Level 2	Tribosystem Level 3	Tribosystem Level 4
	Piston assembly	Compression ring/liner Ring grooves Oil control/scraper ring Piston skirt/liner Small-end conn. rod bearing Large-end conn. rod bearing
Diesel engine	Fuel system	Fuel injector needle/body Fuel injector nozzle (erosion) Turbocharger (nozzle vanes)
	Valve train	Valve stem/valve guide Valve/seats Cam lobes/roller followers Bucket lifters
	Exhaust system	Exhaust gas recirculator (EGR) wastegate bushings

Figure 1.1 An example of how a large tribosystem can be broken down into sublevels containing smaller tribosystems. Level 1, not shown in the figure, is the entire vehicle, the driver, and its environment.

322 [3] has proposed six levels of tribotesting, ranging from full-scale machinery operating in the field to small coupons in bench-scale experiments. As will be discussed later in this book, the effectiveness of a simulation in obtaining meaningful data to screen or select materials or lubricants in practical tribology problem solving depends on matching the key characteristics of the engineering tribosystem with those of the test bench, which is its own tribosystem.

When diagnosing wear problems, especially those concerning lubricants, tribosystems can be characterized as being either open or closed [4]. Open systems have the potential to introduce contaminants or chemical species into the tribosystem while operating. For example, jaw crushers in ore processing or recycling plants are continually being fed new rock, and the composition of the foreign bodies in the input stream may contain unexpected wear-causing materials, like nuts, bolts, or spalled wear plates broken off upstream from the mining equipment itself. Likewise, rolling mill rolls could pick up mill scale or hard particles from elsewhere in the plant.

A transmission fluid pump with a recirculating working fluid and a bearing that is "sealed for life" are examples of closed tribosystems. Some systems have both open and closed characteristics. Internal combustion engines contain recirculating oil and coolant systems, but there is a potential to introduce wear-causing material through the fuel or air intake if filters are not working properly. Obviously, wear debris particles can be generated in both open and closed tribosystems. Years ago, an automotive engine manufacturer had postmanufacturing problems with leftover casting sand from the cylinder blocks finding its way into their engines. Here, an external source of abradants produced wear in what is designed to be a closed tribosystem.

The primary goal of this book is to present a systematic approach for defining tribosystems and their characteristics. In that way, root cause analyses and tribological problem solving can be facilitated, as can the selection of materials, lubricants, and test methods for use in evaluating candidate solutions. While the flexibility of the approach makes it useful for attacking both basic and applied tribology problems, the emphasis in this book is mainly on engineering challenges.

Having established the concept of a tribosystem, the rest of this book is focused on how that basic concept can be applied to analyze and diagnose wear problems. Chapter 2 describes how wear problems present themselves. It includes a discussion of how wear problems are detected, quantified, and monitored. Chapter 3 presents the author's approach to categorizing wear or surface damage based on the type of relative motion that is producing them and their observed features or artifacts. It discusses the role of terminology in the process, including both formal definitions and field-specific jargon, and it introduces a coding system that can provide a measure of consistency and a shorthand system for identifying the type(s) of wear. Since visual observations, both unaided and aided, are key in the diagnosis of wear and surface damage, Chapter 4 highlights a range of surface imaging tools, specimen preparation methods, and their practical advantages and shortcomings. Chapter 5 describes the tribosystem analysis form and how tribosystems, even relatively complex ones, can be characterized. Chapter 6 adds a discussion of problem-solving options, the selection and types of information available from wear testing, and a few examples of matching the performance of large-scale tribosystems with data from simulative test methods.

Finding solutions to wear problems is not equivalent to solving friction problems, but a tribosystem analysis approach can be applied to define friction-critical situations as well. (Recall that the word *tribology* has its roots in the Greek word for rubbing with friction.) As will be discussed later, friction is a manifestation of the energy available to do work between rubbing materials, and creating wear particles is one of the ways (but not the only way) in which that energy is dissipated. Therefore, the relationship between friction and wear varies depending on the nature of each tribosystem, and many similar concerns are involved in both friction problem diagnosis and wear problem analysis. At the center of it all is the notion of a properly functioning tribosystem and how the knowledge of its details enables one to improve its performance.

References

1. ASTM G40-15 (2015) "Terminology relating to wear and erosion," in *ASTM Annual Book of Standards*, Vol. 03.02, ASTM International, West Conshohocken, PA, p. 164, http://www.astm.org/BOOKSTORE/.
2. R. G. Bayer (2002) *Wear Analysis for Engineers*, HNB Publishing, New York, Chapter 3.1.
3. "Kategorien der Verschleißprüfung," Standard DIN 50 322, Deutsche Inst. fur Normen, Berlin, Germany.
4. P. J. Blau (1989) *Friction and Wear Transitions of Materials: Break-in, Run-in, Wear-in*, Noyes Publications, Saddle River, NJ.

Chapter 2

How Wear Problems Reveal Themselves

Tribosystem analysis (TSA) is basically a process to dissect and organize the mechanical, chemical, thermal, and materials-related aspects of a wear-related and/or friction-related condition. Each tribosystem analysis for wear should therefore begin by asking the fundamental question: "How do I know I have a wear problem?" The answer to that question might seem obvious at first, but the ability to detect early indications of wear depends on where and how the problem is occurring.

Superficial indications of wear, like tire wear patterns or marred paint finishes, are easily seen, but in many practical cases, the first indications of a wear problem lie deep inside a machine. They are more subtle and, for all practical purposes, impossible to detect until more serious problems occur downstream. This chapter overviews the kinds of indications that reveal a developing wear problem. These involve the human senses of sight, hearing, smell, and touch. Even taste can be involved if lubricant or wear debris finds its way into food products or containers. In one sense, the various instruments and sensors used in machinery health diagnosis and wear analysis are an extension of the five senses.

There are both direct and indirect indications that wear is occurring. There are also quantitative and qualitative indications of wear. Some wear problems involve cosmetic indications because the function of some products includes how they look to the user. For example, a great deal of clothing is discarded because it looks worn, not because it fails to provide body coverage or warmth. Likewise, some machine components are replaced because they appear to be worn, not because they would not continue to function "mechanically" for a while longer. A disreputable auto mechanic might show you a costly transmission part from your car and claim that it is so worn that it should be replaced. Unless you have worked on automobiles and have experience in this area, few consumers have the experience to dispute such a claim. The need to do something about wear then becomes a matter of subjective judgment or trust in an "expert opinion." We all know that this can be a costly decision and we tend to err on the conservative side if we can afford it.

There are many possible causes for excessive wear in a tribosystem, for example,

- Applying loads, speeds, or operating conditions in excess of the recommended design limits
- Loading a component in a direction other than that for which it was designed (say, an axial overload on a radial bearing)
- Loss, "starvation," or degradation of a lubricant
- Failure of a lubricant filtration system to remove damaging contaminants or debris
- Operating a component beyond its statistical "design life"
- External influences such as heat, vibration, or corrosion from the environment
- Transitions in wear mode, such as the onset of scuffing or galling
- Improper assembly or misalignment of moving parts
- Electrical grounding problems leading to spark damage
- Transition of micropitting damage to macropitting and spalling

Sometimes, the proper function of a product depends on its wearing away. For example, a pencil is designed to wear the graphite from its tip onto the paper. Making the pencil core very hard or too wear resistant would rip the paper without leaving the desired markings. Some grinding wheels are "self-dressing" because the grits are designed to fracture so as to create fresh cutting points. Some kinds of knives, cutting tools, and shearing blades are "self-sharpening" as well. So-called abradable seals in jet engine turbine shrouds wear so as to properly conform to the turbine blade tips and produce a better gas path seal that raises engine efficiency. Therefore, wear is not always a bad thing if it is controlled. Likewise, friction reduction that can improve the energy efficiency of a machine is not always a good thing. We depend on friction for the traction of running shoes, the life-saving effectiveness of automobile brakes, and the grip of friction clutches. Friction control, not necessarily its minimization, is an objective in many engineering tribosystems.

The foregoing examples illustrate that each tribosystem should be analyzed by the components' intended functions, be they cosmetic or otherwise. Wear tests that are applied in material selection or design should be selected to address targeted functions. It is appropriate to ask not only how we know we have a wear or friction problem but also how serious it is and what consequences would ensue if we did nothing about it.

2.1 Direct and Indirect Indications of Wear and Friction Problems

Broadly, wear problems present themselves in one or more of these four ways:

a. Surface appearance
b. Dimensional changes relative to the unworn state
c. Reduced mechanical functionality or sensory indications (smells, rattles, abnormal heat, etc.)
d. The presence of wear by-products or their effects on other tribosystem components

Appearance and dimensional changes (*a*) and (*b*) are direct indications of wear, while (*c*) and (*d*) are indirect indications. These indications are shown in Figure 2.1. The discovery of one wear indicator may lead the engineer to apply other methods of diagnosis and examination to confirm its presence and to ascertain the magnitude of the problem. An example is the use of oil analysis (OA), which indicates the presence of metallic wear flakes of a certain composition. This in turn leads one to focus on metallic-bearing components that have the same composition as the debris and are likely to experience the kind of wear that the morphology and size of the debris from the oil analysis suggests.

Direct indications are those that involve either quantitative changes in dimensions or visual clues from examining the contacting surfaces in situ, using an inspection technique, or ex situ, after disassembly. *Indirect indications* are abnormal changes

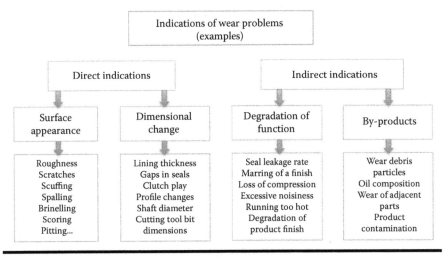

Figure 2.1   A presentation of wear problems with a few examples.

in the nominal operating conditions of a machine or device that leads one to the conclusion that there is a wear or friction problem. Diagnostic tools such as vibration monitoring (VM), oil analysis, or motor current signal analysis can be used to indicate wear indirectly. A few examples of indications of wear problems are listed in Table 2.1.

A practical example of direct, quantitative indications is the guidance provided at a commercial tire service center. Three types of wear indications are used for diagnosing wear in automotive tires: tread depth gauges, embedded wear indicators, and wear pattern charts. The former two indicate the remaining tread life and the latter provides a guide to possible causes for abnormal tire wear. Some manufacturers provide credit-card-sized paper gauges that can be inserted into the treads to indicate the degree of wear. For example, Table 2.2 indicates the degree of wear for two different Michelin tire types [1]. Since 1968, the U.S. government has required tire manufacturers to embed wear indicators into the treads. These indicators become visible when the remaining tread thickness is less than 2/32 inches (1.59 mm).

Condition monitoring, whose spectrum of diagnostic tools includes both direct and indirect wear measuring tools, comprises a large body of knowledge and has created an entire industry that provides sensors, automation, and analytical methods to detect wear and friction problems. New sensors and forms of diagnosis continue to appear, so these examples represent only a snapshot of this evolving field.

Regardless of the form of the surface damage and wear that is occurring, it needs to be detected and caught before the function of the component is harmfully compromised. This is particularly relevant to predictive maintenance programs. In situations where machinery is not run in a consistent or continuous state, but rather is used only periodically, the maintenance interval may need to be based not on the presumption of a constant timeline but rather on a usage-based maintenance interval [2]. This concept has been developed and applied, for example, in the case of sand abrasion of military hardware [3]. The following sections highlight the principles of common methods of wear and surface damage detection.

2.2 Visual Inspection and Surface Imaging

Visual inspection and surface imaging are among the most common ways to detect and assess the relative severity of wear problems, but sometimes, the worn surface is not visible without disassembly of the problem component. While naked-eye examination alone is extremely useful in many cases, enhanced magnification and image processing techniques offer further insights. Because surface imaging and feature measurement is such an important area of wear diagnosis, Chapter 4 is devoted to it. Suffice it to say that the magnitude of enlargement, illumination conditions, sources of image contrast (light optics versus electron optics or acoustics), and ability to measure features varies between methods. Thus, they should be

Table 2.1   Indications of Wear Problems—Some Examples

Tribosystem	Qualitative Observations	Quantitative Measurements
Rolling element bearings	Rattling, vibration, grinding noises, irregular rate of rotation, looseness of the balls or rollers in their cages, pits and spalls in disassembled bearings	Dimensional changes in the balls or races, increase in surface roughness, increased density of micropits per unit area of contact
Ball mill balls or linings	Observations of ball damage, ball or lining materials mixed in with the product, reduced output of ground material	Ball size reduction, electrical power monitoring, ball mill liner wear measurements, distribution of product sizes
Human heart valve leakage	Murmurs, fatigue, shortness of breath, swelling of ankles, dizziness	Echocardiogram, radiography imaging, blood pressure, other medical procedures
Brake (car or truck)	More pressure to stop, noise from a wear indicator, scoring of the counterface, excessive dust	Lining thickness, stopping distance increases
Cutting tool, machining	Increase in cutting force or chatter; damaged workpiece surface; uncharacteristic noise; general worn appearance of the nose, rake face, and clearance face; chip appearance	Number of parts produced up to wear-out, measurable noise or vibration sensor output, surface finish of machined surfaces degrades or becomes nonuniform
Submarine shaft seals	Excess water inboard, lower turning torque, boat trim	Increased pumping rate on the sump
Exterior solar panels	Reduction of electrical output, appearance of erosion or abrasion on the exposed surfaces	Measurable reduction in output, change in transmittance of the outer surfaces
Motors and gearboxes	Color of lubricant, unusual smells, leaks around seals, dripping on the floor, filter appearance, and replacement frequency	Lubricant analysis, including composition and quantities such as acid or base numbers, percentage metals, and more

(Continued)

Table 2.1 (Continued)   Indications of Wear Problems—Some Examples

Tribosystem	Qualitative Observations	Quantitative Measurements
Sheet metal forming rolls	Product defects and deterioration of surface finish	Surface roughness increases
Internal combustion engines	Loss of compression, burning fuel or smoking oil, rattles or loss of responsiveness, more frequent oil addition, oil color, dripping fluids	Loss of cylinder pressure, loss in torque, oil analysis, oil consumption rate
Diamond single-point turning tool	Surface quality of the workpiece	Reflectivity based from the workpiece surface, correlated with tool tip shape
Vehicle tires	Noise, vibrations, loss of traction, worn appearance, and wear patterns	Tread depth, appearance of wear indicators built into the tread

Table 2.2   Indications of Tire Wear from Tread Depth

Wear Condition	Normal Tire Models, Fractional Inches (mm)	Premier® A/S Model, Fractional Inches (mm)
New	10/32 (7.94)	8.5/32 (6.75)
Half-worn	6/32 (4.76)	5/32 (3.97)
Replace soon	4/32 (3.18)	3/32 (2.38)
Replace now	2/32 (1.59)	2/32 (1.59)

Source: *Introducing the Michelin Premier® A/S Tire*, Michelin North America, Greenville, SC, 2014 [1].

selected based on the type of surfaces to be examined. In fact, the use of multiple imaging methods that make up for technique limitations can be complementary, providing more insight than using one method alone.

2.3 Vibration Monitoring and Acoustic Emissions

In the days before the introduction of extensive electronic sensor systems for engine diagnosis, automotive technicians would sometimes use a doctor's stethoscope to listen for the internal heartbeat of moving parts. Many a worn bearing or loose seal or gear was discovered in this way, but it required training and experience. In addition, moving mechanical assemblies containing frictional interfaces can generate noise and vibrations. Sensors, like accelerometers, can therefore be placed in critical areas and their signals can be compared to well-running and problematic conditions. Depending on the sophistication of the reference signal databases, a comparison of vibration spectra can help to track down the location, nature, and severity of internal damage. VM is a key for monitoring the state of wear or friction in locations that are inaccessible or otherwise difficult to inspect.

In a handbook article on diagnostics [4], Cowan defines *vibration* generally as "periodic motion about an equilibrium position." He mentions some causes of forced vibrations in equipment:

- Unbalance
- Rubbing
- Looseness of components
- Misalignment
- Internal friction, cracking, or resonance

Wear can be a source of vibrations, but it is not the only cause. Other causes include corrosion damage, external contaminants, assembly misalignment, part machining errors, and even missing parts. Transducers and signal processing systems of many kinds are available depending on the application and environment of use, so a detailed treatment of VM is beyond the scope of this book. Generally, transducers are used to capture a reference signal over a period of time, analyze the spectrum of frequencies or the signal magnitude at a given frequency, and observe any changes or departures from normal operation. Lower frequencies (say, 10 Hz to 1500 Hz) are used for absolute velocity measurements, and higher frequencies of up to about 20,000 Hz are used for more detailed diagnosis.

Bearings and gear trains are popular areas for vibration monitoring. An example of the application of VM to rolling element bearings was provided by Lacey [5] in a short review article of the subject. Lacey discusses signal capture and how the characteristics of the spectra can be interpreted. In one of his sample cases, Lacey discusses the signals generated by a tapered roller bearing with a 432-mm bore diameter operating at 394 rev/min. The shaft was driven at 936 rev/min and reduced by gearing. The mesh frequency was 374.4 Hz. Table 2.3 summarizes Lacey's analysis of the captured spectrum (see the reference for additional details). The presence of the various harmonics of 6.56 Hz suggests deterioration of one or more rollers, as was confirmed in later examination. Substandard initial bearing quality and machining defects can also lead to vibrations.

While VM is used to detect problems in numerous engineering tribosystems, other sources of sound besides periodic vibrations can help diagnose wear and friction issues as well. In fact, a large number of case studies have been published in the area of VM and, to a lesser extent, acoustic emissions (AEs), including some which

Table 2.3 Indications of Roller Deterioration in a Bearing

Frequency Peak (Hz)	Possible Cause or Interpretation
2.93	Bearing cage speed—indicates roller damage
6.56	Corresponds to the shaft rotation—indicates possible roller damage
62.4	Second harmonic of roller rotational frequency—roller deterioration indication
186.5	Third harmonic of roller rotational frequency—roller deterioration indication
374.4	Gear mesh frequency
497	8th harmonic of roller rotational frequency—roller deterioration indication
560	9th harmonic of roller rotational frequency—roller deterioration indication
748	12th harmonic of roller rotational frequency—roller deterioration indication
873	14th harmonic of roller rotational frequency—roller deterioration indication
936	15th harmonic of roller rotational frequency—roller deterioration indication

Source: Lacey, S., *Maintenance Asset Manag.,* 23, 32–42, 2008 [5].

are focused more on basic research into sliding phenomena than on applications and failure detection (e.g., Benabdallah and Aguilar [6], Lingard et al. [7], Hase et al. [8,9]). Rubbing, fracture, impact, or grinding action can all generate acoustic emissions within materials. Listening for this type of acoustic emission can help one to detect damage in buried interfaces. Similar to VM and accelerometers, AE sensors (akin to microphones) can be used to probe interior interfaces, but the signal for such events must be strong and distinct enough to find its way to the sensor. Therefore, one challenge with this approach is uniquely distinguishing the signals arising from wear and surface damage from the background noise generated during the normal operation of a system (e.g., pumps, fluid flows, structural vibrations, mechanical switches, thermal expansion noises, seal squeals). One need only envision the hundreds of contacting surfaces in a gearbox or transmission systems containing bearings, gears, and seals to soberly understand how difficult it could be to pinpoint both the location and severity of wear within such mechanical assemblies. Furthermore, what are the chances that the sensors, which sometimes sample data periodically and not continuously, would be actively monitoring the tribosystem just at the instant that a fracture or spall happens to occur?

2.4 Motor Current Signature Analysis

Motor current signature analysis (MCSA), also called motor current analysis, is a method to externally probe the mechanical workings of a motor-driven assembly by monitoring the current that the motor draws. Induction motors are one common application. Also, in a motor-driven valve in fluid or material processing systems, the power draw may rise if the valve becomes stuck or corroded so more current is required to operate it. The key is to compare the sampled current profiles with those expected for healthy equipment.

The basic principle of the method is to record the current drawn in the time domain and use a fast Fourier transform or other mathematical techniques like wavelet transforms to analyze the resulting spectra [10–13]. Singh et al. [14] applied the basic principles of MCSA to identify and diagnose problems with rolling element bearings and to compare the information obtained with MCSA with that obtainable from VM (see Section 2.3). Wavelet transforms were used as a mathematical processing tool to conduct multiscale analysis of the signals and transients embedded in them.

Demodulation of a MCSA can also be used in processing to subtract out the carrier frequency (e.g., 60 Hz in the United States) and other characteristic internal operating frequencies from the raw spectrum to enable the resulting signal to act as an indicator of changes or trends in bearing performance. However, the interpretation of the information from techniques like VM and MCSA can require considerable expertise, as well as a database of different failure modes for the particular component of interest. Further applications may be found on websites, for example, see Fossum [13].

In summary, it may be tempting to base a wear analysis on only one diagnostic technique like MCSA ("If you have a hammer, you tend to see every problem as a nail."—Abraham Maslow), but it is impossible to claim, without having examined every possible cause, that a specific signal anomaly *uniquely* indicates that a certain type of wear condition is occurring. A confirming diagnosis by one or more additional techniques and direct examination is advisable.

2.5 Oil Analysis

Oil analysis (OA) is an extensive subject whose focus can be subdivided into (*a*) fluid composition and thermophysical properties (e.g., room temperature viscosity, volatility, thermal conductivity, temperature-viscosity coefficient), (*b*) external contamination level, and (*c*) tribosystem-generated wear debris. More extensive discussions of OA, including glossaries of the specialized terminology used in the field, may be found in other publications, for example, Troyer and Fitch [15] and Henderson and May [16]. As a tool of TSA, oil analysis can help to indicate how much wear is occurring, from where the debris is originating, and even what form of wear is taking place. In the current context, it is useful to understand the kinds of information provided by various OA techniques because such information can supplement TSA. In fact, OA results can be included in the TSA form, as discussed in Chapter 5.

The subject of OA includes how to take samples to avoid errors and bias in the results, when to take samples, how much of a sample to take, how to store the samples, what methods to use to analyze the samples, and how to interpret the results of that analysis. All of these aspects can affect the use of OA in the definition of a tribosystem's normal and abnormal states of operation. A lot of useful information about wear can be obtained from lubricating oil samples, and there are a variety of ways to extract it. Some of the information provided by OA is not directly related to wear analysis but rather relates to other aspects of the oil such as its toxicity, volatility, and pourability. It can also alert the operator to internal corrosion problems or unintended leakage of fluids into a closed lubrication system.

Trending is an important element of oil analysis. By studying periodic oil samples, it is possible to track the progression of wear (including running-in) within a tribosystem in which the individual components may be difficult or impossible to inspect without disassembly. For example, oil analysis from an internal combustion engine may reveal whether wear is occurring in the main bearings, on the piston skirts, or piston rings. It may indicate whether the wear is abrasive in nature (cutting chip like debris) or more adhesive (metallic flakes). Therefore, OA can indicate both the locations and types of wear in a tribosystem.

OA can be done as a matter of routine preventive maintenance or as-needed to investigate a suspected wear problem. If there is a sufficient continuing need, OA laboratories can be established on-site. Military equipment maintenance depots (for example, those that service helicopters or fixed wing aircraft) might have on-site OA laboratories. Alternatively, collected samples can be sent to a third-party laboratory that specializes in OA. Such organizations usually provide sampling kits and collection instructions.

2.5.1 Sampling

The method of sampling can bias the results of an OA. Troyer and Fitch [15] list some ways in which misleading or incorrect data can be obtained from bad sampling techniques:

- Sampling under "cold conditions"—not running as usual
- Drain port or vacuum drop tube sampling
- Use of inconsistent sampling methods and locations
- Use of contaminated equipment to take samples

- Insufficient flushing
- Sampling immediately after changing the oil
- Cross-contamination between two different samples
- Waiting too long between sampling and analysis

Even after sampling, handling issues can affect the accuracy of OA results. Foley [17] mentions the following potential sources of sampling and handling errors:

- Failure to use sampling ports and valves where practicable
- Taking cold samples that are not typical of warmed-up equipment
- Incorrect labeling
- Incomplete sample labeling
- Omitting useful observations (such as odors, abnormal heat) when taking samples

OA is broadly divided into chemical analysis of oils (Section 2.5.2) and debris particle (solids) characterization (Sections 2.5.3 and 2.5.4). Owing to unique differences compared to liquid lubricants, there are special methods for the analysis of greases, which, by definition, contain oil and a thickener such as soap. Likewise, methods for particle analysis are different for solid lubricants like graphite powder, silver, molybdenum disulfide, and polytetrafluoroethylene (Teflon®). Still, debris particles can sometimes be collected in these types of situations, and their examination can be helpful even though the quantitative measurement of debris concentration (c.f., in oils) is impractical when solid lubricating films, self-lubricating composites, or powder lubricants are used.

2.5.2 Chemical Analysis of Lubricants and Working Fluids

Various methods are used for the chemical analysis of oils and other working fluids, like hydraulic fluids. A common one is called "spectroscopic oil analysis." The so-called Spectroscopic Oil Analysis Program (SOAP) has become a staple in the field of equipment maintenance and is sometimes coupled with debris separation techniques, like ferrography described in Section 2.5.4, to provide a comprehensive picture of the state of health of key rotating machinery like helicopter transmissions and wind turbine gearboxes.

Sometimes, simply dropping a sample of used oil onto a piece of filter paper and watching the fluid and solids separate can be a useful rudimentary oil analysis. This method has evolved into a standard ASTM D7899 [18]. Like tribosystem analysis in general, it is useful to ask the basic question: "Why do I need these data and what specific information am I looking for?" If certain key identifying compounds or species are known to pinpoint a problem component, then a complete chemical analysis may not be required in order to track the problem. Only one or two witness chemical elements may need to be tracked. On the other hand, additional information, beyond the usual suite of parameters, might be required. For example, in the case of diesel engines, it is helpful to conduct an analysis of the soot content drain oil [19]. If the soot content is more than a few percent, special techniques of separation and analysis may be required in order to achieve a more accurate quantitative result. Therefore, any OA should be tailored to the specific case.

Table 2.4 lists a number of common oil analysis tests and the kinds of information they can provide. References are footnoted in the table.

2.5.3 Debris Particle Characterization

Wear debris, like other solid particles, has a number of features, some of which can be extremely helpful in wear analysis. The features of debris particles carried along by lubricants can change over time, and that is why periodic sampling of lubricants and working fluids can be particularly informative:

- Size or size distribution
- Shape (aspect ratio, roundness, angularity, curvature)
- Composition
- Physical properties (hardness, electrical conductivity)
- Surface features
- Concentration (in a fluid or grease)
- Magnetic character

Glaeser [20] schematically depicted the progression of sizes of metallic wear debris and the severity of wear based on OA. Figure 2.2 is adapted from that reference and shows typical particle size ranges. Note that there is no minimum particle size, even at short times when the wear is said to be "benign," but rather the distribution is truncated at 0.1. Also, as might be expected, the more severe the wear, the larger the concentration of larger particles. While the schematic portrayal in Figure 2.2 suggests that wear steadily increases with time, in some cases, the rate of debris generation can accelerate (during wear-out) or decelerate (after run-in), and it is risky to extrapolate the debris concentration or the time until wear-out by simply assuming a linear dependence of concentration or size on the time of machine operation.

Wear debris can be captured in a variety of ways. Dry contact generates a great deal of debris, and in this case, sampling is easier than extracting the particles from oil. Lightly brushing dry wear surfaces with a soft brush or using adhesive tape can be used to collect dry wear debris. In laboratory experiments, a scanning electron microscope (SEM) stub wetted with silver paint or adhesive can be held under a sliding contact to capture debris.

Some common methods to extract debris from oil are described in the following section. As mentioned in the introduction to Section 2.5, the method of collecting samples can bias the results of a wear analysis, and users should be aware of potential limitations and biases associated with any of the techniques.

Table 2.4   Oil Analysis Tests and the Type of Information They Can Provide

Name of Technique	Table Ref.	Description	Typical Information
Blotter spot test	i	Place an oil drop on a filter paper strip, allow to dry, suspend strip horizontally, bend one end down and dip it into *n*-heptane in a pan, and note the separation pattern.	Comparison of the pattern on the strip to reference patterns for free carbon, insoluble resins or oxidation products, source of contaminants, fuel in engine oil, glycol in engine oil.
Crackle test	ii, iii	Set a hotplate to 160°C and observe bubble formation or "crackling" within a drop of oil; other versions of this test are used for greases.	Bubble size, if any, is a rough indication of water content in an oil sample.
Elemental spectroscopy	iv	Detects the concentration and presence of 15 or more elements in the oil.	Composition indicates any elements that should or should not be present in the oil, due to contamination.
Ferrographic analysis	iv	Magnetically separates particles from a small oil sample for imaging or counting.	Sizes and shapes of ferrous and nonferrous wear particles in a sample, type of wear present.
Flash point test	iv	Heats oil samples gradually under an open flame until ignition occurs or until reaching a temperature limit for pass/fail.	Presence of volatile compounds, presence of fuel contamination in engine oil.

(Continued)

Table 2.4 (Continued) Oil Analysis Tests and the Type of Information They Can Provide

Name of Technique	Table Ref.	Description	Typical Information
Fourier transfer infrared spectroscopy (FTIR)	iv	Produces a spectrum of various compounds in the oil.	Constituents of an oil sample, including compounds that should and should not (contaminants) be present; lubricant "health."
Karl Fischer test	iv	Uses a crackle test (above) or a FTIR test.	Water content in the oil from sources like cooler leakage and moisture condensation.
Oil detection paper	v	Immerse the paper or press it into a soil sample; changes color in the presence of hydrocarbons such as oils, gas, mineral spirits.	Tests of contamination of water or solid for the presence of oil or hydrocarbons.
Magnetic flux analysis	ii	Sample of oil subjected to a magnetic field.	Number and size of particles.
Particle counting	iv	Any of several methods to measure concentration of particles in given size ranges.	Problems with external contamination, filter failure, wear particle size distribution.
Spectrometric analysis	ii	High heat vaporization of metals.	Can detect parts per million of elements.
Total acid number	iv	Milligrams of potassium hydroxide (KOH) needed to neutralize the amount of acid in 1 g of oil.	Oil oxidation, depletion of additives, corrosiveness, incorrect oil composition for specified application.

(Continued)

Table 2.4 (Continued) Oil Analysis Tests and the Type of Information They Can Provide

Name of Technique	Table Ref.	Description	Typical Information
Total base number	iv	Milligrams of KOH per gram of oil, typically in a crankcase.	Reserve alkalinity available to neutralize any acid that forms, can indicate a number of problems like soot contamination, poor combustion, blow-by in an engine.
Viscosity	iv	Any of a number of techniques to measure flow resistance of the oil.	Chemical degradation of the oil, excessive contamination, volatilization of oil components, mixing or top-off with other oils.

Sources: (i) http://www.machinerylubrication.com and ASTM D-7899-13, *Standard Test Method for Measuring the Merit of Dispersancy of In-Service Engine Oils with Blotter Spot Method,* ASTM International, W. Conshohocken, PA [18]; (ii) http://www.machinerylubrication.com; (iii) http://www.greasology.org /water_test.htm; (iv) Troyer, D. and Fitch, J., *Oil Analysis Basics,* Noria Corp., Tulsa, OK, 1999 [15]; (v) http://www.newpig.com.

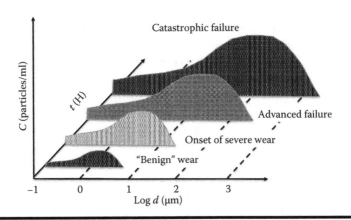

Figure 2.2 Shifting distributions of debris particle sizes (*d*) and concentrations (*C*) as a function of time (*t*). (Adapted from Glaeser, W.A., Wear debris classification, in Bhushan B., ed., *Modern Tribology Handbook,* Vol. 1, CRC Press/ Taylor & Francis, Boca Raton, FL, 2001 [20].)

One method to quantify solid particle concentrations in contaminated oil may be found in ISO 4406. The standard lists a series of size ranges and their corresponding codes [21]. Concentration is given in units of particles per milliliter. A close estimate of the minimum concentration (C_{min}) and maximum (C_{max}) concentrations of particles that fall within a given ISO range number (R) can be found from

$$\log C_{min} = 0.30(R) - 1.996 \tag{2.1}$$

and

$$\log C_{max} = 0.30(R) - 2.297. \tag{2.2}$$

Therefore, for $R = 14$, the particle concentrations would range between 80 and 160 per milliliter of oil. Subscripts, referring to the mean particle size in micrometers, can be used with range numbers to benchmark typical or abnormal distributions. The engineer can select say two or three ranges within sample to use for this purpose. For example, R_5 would be the range number for 5-μm particles, and these can be listed in series to indicate the size ranges for that size particles. An R_5/R_{10} of 14/10 would mean that the concentration level for 5- and 10-μm particles would be represented by range numbers 14 and 10, respectively. A discussion of this approach is given in an article by Gresham [22] and on the Internet [23].

2.5.4 Filtration, Magnetic Traps, and Ferrography

Effective filtration can be the key to ensuring a long component wear life in liquid lubricated tribosystems. However, because particles are retained in filters or specialized devices attached to filtration systems, they can also be a means not only for cleaning but also for augmenting tribosystem analysis. Methods to obtain samples of oil and the wear debris particles they contain range from simple dipstick sampling and extracting particles from oil filters to the use of removable magnetic plugs or inline detectors that are mounted in a bypass circuit designed into an oil recirculation system. Filtration and magnetic traps are both extensive subjects in themselves and will not be covered here. Instead, the reader is referred to excellent reviews presented elsewhere [24,25]. In addition, there is an extensive, online image catalog containing images of wear particles of various metals with captions to describe the sources of the samples [26]. For transparent or translucent materials, both transmitted light and electron microscope images are provided in the catalog.

Ferrography is an established method for wear debris separation and analysis in which a small sample of oil is allowed to drain down a ramp formed by a glass slide that is placed directly above a magnetic field (see Figure 2.3). Development of the method is credited to Bowen and Wescott in the 1970s (e.g., Ref. [27]). The

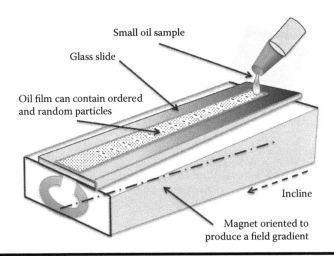

Small oil sample

Glass slide

Oil film can contain ordered
and random particles

Incline

Magnet oriented to
produce a field gradient

Figure 2.3   A schematic diagram of the ferrographic method introduced in the 1970s.

magnetic field is designed to be stronger at one end of the slide than the other so that drifting particles are separated based on their relative magnetic response to the field strength. When the oil is so treated to form chains of particles along the length of the slide, the resulting arrangement is termed a *ferrogram* and is inspected under a microscope. Some ferrographs are called "direct reading" because the ferrograms can be illuminated by a light source above and the signal can be detected by photodiodes beneath it. By using more than one photodiode, a range of particle sizes as large as 0.1 to 300 µm or greater can be sorted. Magnetic particles tend to group and separate along the field lines, but nonmagnetic particles are also detectable since they tend to stand out from the orderly magnetic particles. Like other forms of microscopy, ferrograms can also provide clues to the type of wear being produced and its likely location. They can be illuminated by reflected light, transmitted light, or polarized transmitted light to help sort out the kinds of particles and their sources. In addition, the slides can be heat treated (e.g., a few minutes at 315°C) or tempered to aid in compositional analysis [28]. Larger particles are associated with more severe wear. Some users target a particle size of over 30 µm as equating to "severe" or "abnormal wear."

Sample ferrograms produced at a commercial wear analysis laboratory are given in Figure 2.4a–d. According to the source, Wear Check Canada International (Burlington, Ontario, Canada, 2015), the ferrogram in Figure 2.4a shows "severe cutting wear particles (curled chips) from a pulverizer gearbox (Raymond 783 RPS) in a thermal power station. The dark particles in the ferrogram are most likely abrasive coal particles as this unit pulverizes coal for the boiler furnace. So, in this case these would be 3-body abrasive particles (caused by an abrasive contaminant) as opposed to misalignment." The ferrogram in Figure 2.4b shows "severe

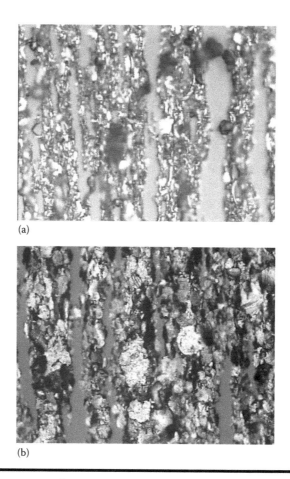

(a)

(b)

Figure 2.4 **(a) Severe abrasive wear particles; (b) adhesive wear and contact fatigue wear particles.** *(Continued)*

combined sliding and rolling fatigue particles from a Dodge TDT 825 reduction gearbox. Combined sliding and rolling fatigue is common from gearboxes. The rolling fatigue particles are from the pitch line of the gear tooth as the two teeth roll past each other. The remainder of the tooth generates sliding wear particles during abnormal loading/poor lubrication film." The ferrogram in Figure 2.4c is a "typical ferrogram from a gearbox showing no significant wear (normal). There are some potash particles present in the ferrogram since this gearbox is in a potash mill." The ferrogram in Figure 2.4d is a "typical normal ferrogram from a turbine thrust bearing from a hydroelectric power plant." The use of ferrography is common in predictive maintenance programs in military, aerospace, and industrial gearboxes and other closed tribosystems in which the lubricant circulates and can be periodically sampled for analysis.

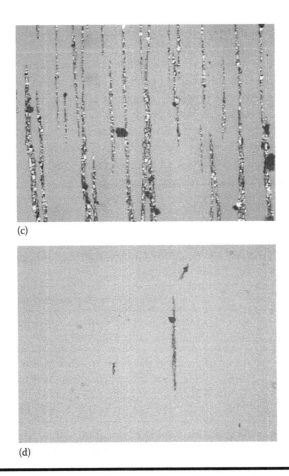

(c)

(d)

Figure 2.4 (Continued) Ferrograms showing normal wear from (c) a gearbox and (d) a turbine thrust bearing. All images were acquired at ×200 using bichromatic light: reflected white light plus transmitted green light. (Ferrograms and their descriptions are provided courtesy of Bill Quesnel, Wear Check Canada International, Burlington, Ontario, Canada, 2015. Additional examples may be found online at: http://lubrication-management.com/en/services/lubricating-oil -analysis/.)

According to Barrett and McMahon [29], six categories of particles can be differentiated from analytical ferrograms:

1. White nonferrous (such as aluminum or chromium, randomly oriented)
2. Copper (bright yellow)
3. Babbitt (Pb–Sn, appearing as gray or with red or blue spots)
4. Contaminants (e.g., silica and nonmetallic dirt)
5. Fibers (from filters or from external sources)
6. Ferrous particles including oxides

The last category, ferrous, can be differentiated as high-alloy steel, low-alloy steel, dark oxides, cast iron, and red oxides.

Special particle analysis procedures have been developed for greases. Since grease is composed of oil and a thickening agent (i.e., a two-phase mixture), it exhibits certain characteristics of flow and interfacial retention that are not found in liquid lubricants. Particles like wear debris also move differently or agglomerate when entrained in greases as compared to oils (see an example of wear particles in grease in Chapter 4). Therefore, standards have been developed in the lubricants industry to analyze particles collected from grease-lubricated tribosystems. An example of these methods, selected from Kaperick [30], is provided in Table 2.5.

Table 2.5 Standardized Methods for Grease Particle Collection and Analysis

Document	Short Description of Method
DIN 51825	Provides rating based on particles >25 μm in a kilogram of grease, as measured in DIN 51813.
DIN 51813	0.5 kg of grease is passed through a 25-μm mesh, residues are collected and dried, and grease is dissolved. Remainder is filtered, and >25-μm particles are weighed. "Pass" is defined as <20 mg particles/kg grease.
Def Stan 05-50 (Part 39)	Graphite content in grease. Boil with solvent reflux, filter, then weigh retained solids to obtain percentage as 100 × (weight of solids/weight of sample).
MIL-G-81322	Used for aircraft grease. Pass = less than 1000 particles (25–74 μm)/cm^3 and none >75 μm.
MIL-G-81937	Used for clean instrumentation grease. Pass = less than 1000 particles (10–34 μm)/cm^3 and none >35 μm.
FTM 791 Method 3005.4	Residual contamination; 1 cm^3 grease on a rectangular template supported between glass slides.
ASTM D1404	Deleterious particles in lubricating grease. Count arch-shaped particles in a 0.25 cm^3 sample of grease squeezed between acrylic plates at 200 psi and rotated 30 degrees.

Source: Kaperick, J., *Tribol. Lubr. Tech.*, 71, 32–36, 2015 [30].

Note: ASTM, ASTM International Standard designation; Def Stan, UK Ministry of Defense Standard; DIN, German standard; FTM, U.S. Federal Test Method; MIL, U.S. Military Standard.

2.6 Radionuclide Wear Detection

Since about 1949, radionuclides have been successfully used to monitor the wear of internal, noninspectable parts, like internal combustion engine piston rings and bearings, a technique that became more widely used in the 1960s. A refinement of this method is called *surface layer activation* (SLA) [31,32]. The principle is to irradiate the parts of interest using a source of neutrons or energetic ions with a convenient half-life and obtain a calibration curve of radiation versus time for the nonworn part. Gamma ray detectors are placed outside the tribosystem and radiation levels are compared with the baseline decay rate. Lubricant samples can also be examined to reveal the concentration of activated species carried into the oil. Like any radioactivity-based method, there are issues of facility certification, safety requirements, and training in radionuclide handling. From a timing standpoint, there is a need to schedule and conduct experiments efficiently since the half-life of the typical tracer isotopes used for thin layer activation of wear surfaces is usually measured in hours or days once a part has been treated. Installation into wear test components and running of tests need to be scheduled promptly.

Note that the SLA method of wear detection implicitly assumes that the wear loss of material detected this way is uniformly distributed over the activated area; however, if wear is localized, such as the removal of a small area of pits or fatigue spalls, and if the remainder of the layer were to remain intact, the degree of wear (depth) in certain spots might be larger than presumed from the averaged readings. Confirmatory post hoc analysis is therefore advisable to see if the assumption of uniform wear loss is justified.

SLA continues to be used in recent years despite a tightening of regulations for handling radioactive materials. It has been applied not only to automotive engine components but also to pumps, prosthetics, and aerospace components [33–35].

2.7 Summary

Wear problems present themselves in different ways, and the tools used to diagnose them also vary widely in sophistication, availability, and cost. From a general perspective and considering the many forms of surface damage, visual inspection is probably the most used tool to detect wear, but sensory cues, like abnormal heat or vibrations, or a change in the condition of a lubricant and abnormal forms or amounts of debris collected on oil filters, are some common indications. Some cues are direct and others are more subtle. While quantitative indications, like a change in the debris concentration in an oil or an increase in a seal leakage rate, quantitatively reflect the severity of the wear problem, qualitative indicators can serve as an alert that there is a growing problem. Table 2.6 reviews the tools for wear diagnosis discussed in this chapter. Often, an intelligent combination of diagnostic tools can be used to gain more depth of information and support the diagnosis of a wear

Table 2.6   Wear Detection and Diagnostic Tools

Tool	Description (Book Section)	Book Section or Chapter
Visual examination	Naked eye — unaided Hand lens (loupe, magnifier) Light optical microscopy (4.1) Electron optical microscopy	2.2 and Chapter 4
Thermography	Imaging based on frictional temperatures or surface emissivity (4.4.2)	Chapter 4
Acoustic imaging	Scanning acoustic microscopy (4.4.1)	Chapter 4
Surface roughness	Quantitative light optical imaging Stylus-based and noncontact profilometry (4.2)	Chapter 4
Loss of dimension	Any quantitative method of wear measurement such as surface recession, increasing clearance, leakage	2.1
Mass loss of material	Wear loss of a component after use	2.1
Volume loss of material	Usually inferred from mass loss and density but can be measured using microscopy of wear scar dimensions or computerized 3D image analysis	2.1
Vibration monitoring	Monitoring vibrations using accelerometers	2.3
Motor current analysis	Monitoring the current changes drawn by operating equipment	2.4
Oil analysis	Spectroscopy — various types (2.5.2) Ferrography Blotter spot testing Other: crackle testing, flash point, FTIR, magnetic flux	2.5
Debris particle analysis	Particle count and sizing from liquid samples (2.5.3) Ferrogram analysis of liquids (2.5.4) Grease analysis methods	2.5
Surface layer activation	Change in decrease in activity of worn activated surface layers compared to the reference decay behavior of nonworn areas	2.6

problem (e.g., Peng and Kessissoglou [36]). Therefore, a well-equipped wear diagnostic laboratory will contain imaging and microscopy systems, lubricant analysis testing, surface metrology, mechanical testers (hardness), and experienced technicians to run those instruments.

After detecting the presence of wear or its effects, the next step in a systematic general tribosystem analysis is to identify the dominant type(s) of wear that are present. A hierarchical system of wear categories and a series of shorthand codes to use for reporting them are presented in Chapter 3.

References

1. *Introducing the Michelin Premier® A/S Tire* (2014) Michelin North America, Greenville, SC. Tire dealer sales literature, table of tire wear.
2. T. Tinga (2010) "Application of physical failure models to enable usage and load-based maintenance," *Reliability Engineering & System Safety*, Vol. 95, pp. 1061–1075.
3. M. Woldman, T. Tinga, E. Van Der Heide, and M. A. Masen (2015) "Abrasive wear based predictive maintenance for systems operating in sandy conditions," *Wear*, Vol. 338–339, pp. 316–324.
4. R. S. Cowan (2012) "Diagnostics," in *Handbook of Lubrication and Tribology*, R. W. Bruce, ed., CRC Press/Taylor & Francis, Boca Raton, FL, pp. 65-1–65-13.
5. S. Lacey (2008) "An overview of bearing vibration analysis," *Maintenance and Asset Management*, Vol. 23 (6), pp. 32–42, http://www.maintenanceonline.co.uk/mainte nanceonline/content_images/p32-42%20Lacey%20paper%20M&AM.pdf.
6. H. S. Benabdallah and D. A. Aguilar (2008) "Acoustic emission and its relationship with friction and wear for sliding contact," *Tribology Transactions*, Vol. 51, pp. 738–747.
7. S. Lingard, C. W. Yu, and C. F. Yau (1993) "Sliding wear studies using acoustic emission," *Wear*, Vol. 139, pp. 597–604.
8. A. Hase, M. Wada, and H. Mishina (2008) "Acoustic emission signals and wear phenomena on severe-mild wear transition," *Tribology Online*, Vol. 3 (5), pp. 298–303.
9. A. Hase, M. Wada, and H. Mishina (2006) "Correlation of abrasive wear phenomenon and AE signals," *Japanese Journal of Tribology*, Vol. 51 (10), pp. 752–759.
10. D. M. Eisenberg and H. D. Haynes (1992) "Motor-current signature analysis," in *ASM Handbook, Vol. 18, Friction, Lubrication, and Wear Technology*, P. J. Blau, ed., ASM International, Materials Park, OH, pp. 313–318.
11. X. Li and S. K. Tso (1999) "Drill wear monitoring based on current signals," *Wear*, Vol. 231 (2), pp. 172–178.
12. W. T. Thomson and R. J. Gilmore, "Motor current signal analysis to detect faults in induction motor drives—Fundamentals, data interpretation, and industrial case histories," http://www.turbo-lab.tamu.edu/proc/turboproc/T32/t32-16.pdf.
13. D. Fossum, "Identifying mechanical faults with motor current signature analysis," www.reliableplant.com/Read/28633/motor-current-signature-analysis.
14. S. Singh, A. Kumar, and N. Kumar (2014) "Motor current signature analysis for bearing fault detection in mechanical systems," *Procedia Materials Science*, Vol. 6, pp. 171–177.
15. D. Troyer and J. Fitch (1999) *Oil Analysis Basics*, Noria Corp, Tulsa, OK.

16. K. O. Henderson and C. J. May (2012) "Lubricant properties and characterization," in *Automotive Lubricants and Testing*, S. C. Tung and G. E. Totten, eds., ASTM International, West Conshohocken, PA, and SAE, Warrendale, PA, pp. 47–59.

17. J. Foley (2015) "Complexity in ISFA (in-service fluid analysis), part XXI," *Tribology & Lubrication Technology*, Vol. 71 (3), pp. 77–79.

18. ASTM D7899-13 (2013) "Standard Test Method for Measuring the Merit of Dispersancy of In-Service Engine Oils with Blotter Spot Method," Annual Book of Standards, Vol. 05.05, ASTM International, W. Conshohocken, PA.

19. J. J. Truhan, J. Qu, and P. J. Blau (2005) "A rig test to measure friction and wear of heavy duty diesel engine piston rings and cylinder liners using realistic lubricants," *Tribology International*, Vol. 38 (3), pp. 211–218.

20. W. A. Glaeser (2001) "Wear debris classification," in *Modern Tribology Handbook*, Vol. 1, B. Bhushan, ed., CRC Press/Taylor & Francis, Boca Raton, FL, pp. 301–315.

21. ISO 4406 (1999) "Method for coding the level of contamination by solid particles." Available from ANSI.

22. R. M. Gresham (2015) "Contamination Control," in *Tribology and Lubrication Technology*, pp. 28–30.

23. "Understanding ISO Codes," http://www.hyprofiltration.com/clientuploads/directory/Knowledge/PDFs/ISO%20Intro.pdf (accessed March 29, 2015).

24. B. Fitch (2009) "Applications and benefits of magnetic filtration," http://www.machinerylubrication.com/Read/794/magnetic-filtration (posted September 2005).

25. K. S. Sutherland and G. Chase (2008) *Filters and Filtration Handbook*, 5th ed., Elsevier, Oxford, United Kingdom, pp. 198–202.

26. McCrone Atlas of Microscopic Particles, http://www.mccroneatlas.com.

27. E. R. Bowen, D. Scott, W. W. Seifert, and V. C. Wescott (1976) "Ferrography," *Tribology International*, Vol. 9 (3), pp. 109–115.

28. F. T. Barwell, E. R. Bowen, J. P. Bowen, and V. C. Wescott (1977) "The use of temper colors in ferrography," *Wear*, Vol. 44, pp. 163–171.

29. M. Barrett and M. McMahon, "Analytical ferrography—Make it work for you, insight services," http://machinerylubrication.com/Read/5/analytical ferrography.

30. J. Kaperick (edited by J. Van Rensselar) (2015) "Grease particle evaluation," *Tribology & Lubrication Technology*, Vol. 71 (9), pp. 32–36.

31. C. C. Blatchley (1992) "Radionuclide methods," in *ASM Handbook, Vol. 18, Friction, Lubrication, and Wear Technology*, P. J. Blau, ed., ASM International, Materials Park, OH, pp. 319–329.

32. C. C. Blatchley, "Surface Layer Activation." This brief online description, somewhat dated, can be found at http://www.pittstate.edu/department/physics/faculty/blatchley/surface-layer-activation.dot.

33. D. C. Eberle, C. M. Wall, and M. B. Treuhaft (2005) "Applications of radioactive tracer technology in the measurement of wear and corrosion," *Wear*, Vol. 259 (7–12), pp. 1462–1471.

34. P. Schaaff, W. Hortsmann, M. Dalmiglio, and U. Holzwarth (2006) "A compact fretting device for testing of biomaterials by means of thin layer activation," *Wear*, Vol. 261 (5–6), pp. 527–539.

35. E. A. Andarawis and C. O. Umeh (2008) "Systems for inspection of shrouds," U.S. Patent 7,852,092 B2, a method for wear damage detection on steam turbine abradable coatings.

36. Z. Peng and N. Kessissoglou (2003) "An integrated approach to fault diagnosis in machinery using wear debris and vibration analysis," *Wear*, Vol. 255, pp. 1221–1232.

Chapter 3

Types of Surface Damage and Wear

The diagnosis of wear problems can be made easier by adopting an organized, hierarchical system of surface damage classification. The current approach considers the type of relative motion, examines the features on the contact surfaces, and then confirms those observations with the kind of debris that is produced. Then one can designate the diagnosis using appropriate nomenclature. In truth, the multidisciplinary nature of tribology has led to the inconsistent use of wear and surface damage terminology in reports, articles, handbooks, and even textbooks. Implementing a consistent system of naming wear types would help to better define durability problems, provide consistency in documentation of wear failures, aid in the compilation of root cause databases, and facilitate wear testing and problem solving. In this chapter, the major forms of surface damage and wear are presented along with a proposed coding scheme. These terms can then be used in the tribosystem analysis form described in Chapter 5.

As described in Chapter 2, wear can be detected in a variety of ways, one of the most common being visual examination. The categorization of wear types presented in this chapter is based on a combination of surface appearance, the nature of debris produced, and the context of its occurrence. Rarely does the initial identification of the dominant type of wear require the use of sophisticated instrumentation. As discussed in the next chapter, a good hand lens or a stereo macroscope (10–50×) can often provide sufficiently telling information for a preliminary wear categorization. However, enhanced examination using various topographic imaging methods, debris collection, oil analysis, hardness testing, and surface chemical analysis can provide deeper insights into the details of surface damage and its causes. Indicating the types of wear and surface damage present is one important part of any tribosystem analysis; however, it is not an end point, but rather the means to finding a solution.

Importance of consistent terminology. If people use different terms to describe the same phenomena, confusion results, and common aspects and trends could be missed. Unfortunately, the occurrence of wear in so many different areas of mechanical technology such as gears, seals, friction materials (brakes and clutches), and bearings has led to the development of specialized terminology within those areas, and the same term may have slightly different connotations depending on its context. For example, the chapter on "Wear Failures" in the ASM International handbook on failure analysis and prevention [1] lists the following five types of wear:

- Abrasive wear
- Erosive wear
- Corrosive wear
- Erosion–corrosion
- Surface fatigue

In the *same* handbook, a different chapter, "Failures of Rolling Element Bearings" [2], distinguishes 13 types of wear:

- Brinelling, false
- Brinelling, true
- Electrical pitting
- Flaking
- Fluting
- Fretting
- Indentation
- Skidding
- Scuffing
- Sliding
- Smearing
- Softening
- Spalling

Not to belabor the point, but in that *same* handbook, another article, "Failures of Gears" [3], also lists 13 types of wear, but not the same 13 as previously mentioned, namely:

- Normal wear (polishing in)
- Moderate and destructive wear (overload)
- Scoring
- Interference wear

- Abrasive wear
- Corrosive wear
- Flaking
- Fluting
- Burning
- Surface fatigue
- Initial pitting
- Destructive pitting
- Spalling

Needless to say, there are many other examples of inconsistencies in naming wear in the technology literature. The point is that consistency in describing wear damage is important in a tribosystem analysis, especially if it forms part of an experience-based database. The challenges of consistent wear terminology have been discussed in a bilateral publication from U.S. and Russian perspectives [4]. Appendix 3A contains a discussion of wear nomenclature and a list of popular jargon.

A system for categorizing the types of wear is now presented. The intent is to enable the investigator to organize wear by type of relative motion between bodies and its characteristic features. This approach also includes suggested shorthand codes that can be used to indicate the types of wear observed when conducting a tribosystem analysis.

3.1 Types of Surface Damage

Generally, surface damage can broadly be divided into four causal types (see also Figure 3.1):

1. *Thermal damage*—This includes softening, melting, ablation, sublimation, and other thermal processes, such as quench cracking or heat checking in bearings, that can cause degradation in surface properties and performance.
2. *Chemical reaction or dissolution*—This includes corrosion, dissolution of the surface in a fluid, diffusion of an embrittling species into a surface, and other forms of chemical attack that can alter the composition and otherwise degrade the properties of materials exposed at a surface.
3. *Radiation damage*—This form of damage relates to the atomic scale interaction of the surface material with energetic particles or packets of radiation that create subsurface defects or spall off atoms or clusters of atoms. Technically, radiation damage could be considered a form of erosive wear, but it occurs at the atomic scale, and that mode of damage is beyond the scope of this book.

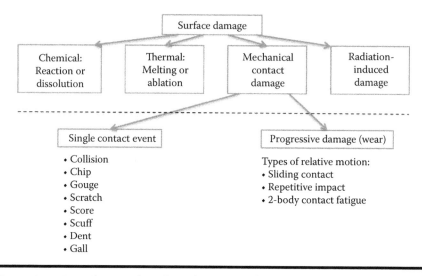

Figure 3.1 Four forms of surface damage and subcategories. When taken as pairs and as combinations of 3 or 4, there are 15 possible interactions. In particular, mechanical damage can occur as single events or progressively.

4. *Mechanical damage*—This form of damage is caused by the mechanical inter-action of a solid surface with another body or bodies so as to alter, deform, or degrade the function or form of the surface or surfaces. Most forms of wear are included in this category, but not all mechanical damage fits the defini-tion of wear that will be presented later.

One could consider wear primarily to be a form of mechanical surface dam-age, but in practice, it may occur in conjunction with other forms of surface damage. For example, when wear of metals occurs in the presence of chemical attack, this interactive process is given the name tribocorrosion. The following examples illustrate some instances where multiple types of surface damage occur together:

■ *Severe abrasive wear during the transport of mineral slurries in ore processing operations*—The basic or acidic character of the slurry can attack the surfaces of rock crushers, conveyors, or piping. Abrasion removes the corrosion prod-ucts (scale) as they form and expose bare metal, producing a synergistic effect. In fact, standard test methods have been developed to quantify this "slurry abrasivity" effect for different minerals and ores.

■ *Wear of dental restoratives*—Wear from chewing occurs in the presence of saliva or chemically active foods.

■ *A gouge in a protective coating on a deep ocean oil drilling rig*—Such a gouge (mechanical damage) can break the surface of a protective coating and initiate corrosion that spreads quickly from that point due to an "anode effect" or simply by crevice corrosion.

■ *Wear of fuel rod claddings in pressurized water nuclear reactors*—Fretting contact of the cladding against a supporting grid, induced by coolant flow, occurs in hot, chemically treated water. Not only does corrosion occur, but also exposure to radiation and hydrogen embrittlement can weaken the metal and accelerate fretting combined with impact wear. (More details may be found in Blau [5].)

The foregoing illustrations depict varying degrees of complexity that can challenge those who attempt to conduct a tribosystem analysis. However, ignoring the potential for interactions between mechanical factors and other forms of surface damage in order to "simplify the problem" could lead to a misidentification of root causes. Sometimes, damage processes can transition from one to another. For example, localized fretting can initiate microcracks that grow under the action of fatigue until they reach a critical crack length that leads to fast fracture. Such a combination of mechanical wear with corrosion has led to hybrid terms like *fretting fatigue*, *fretting corrosion*, and the general term *fretting wear*.

In some cases, one particular process or combination of processes could dominate, and the others, being less important, can be neglected as second-order effects. In recognition of the possibility of synergistic effects and the need to balance costs with material properties, a section in the tribosystem analysis form, to be described in Chapter 5, has been reserved for including additional performance requirements and factors besides tribological ones. These can include corrosion resistance, material availability from suppliers, and costs associated with finishing treatments and basic raw materials.

As shown below the dashed line in Figure 3.1, mechanical damage can be further subdivided into two types: single occurrence and repetitive. The following are common examples of single-occurrence mechanical damage:

■ A scratch on a pair of expensive prescription sunglasses
■ A dent in the front left fender of a 2007 Audi A6 sedan
■ An indentation in the lid of a galvanized steel garbage can
■ A chip on the edge of a floor tile in a restaurant bathroom
■ A gouge on the surface of an 8-pound sledgehammer
■ A firing pin mark on a brass cartridge found at a crime scene
■ Galling and seizure during the tightening of a bolt
■ A key scratch on the door of a restored 1957 Chevrolet convertible
■ A scuff mark on the side of a man's Ferragamo dress shoe
■ A dime-sized clamshell fracture on the windshield of an automobile

Some of the items in this list are considered "normal wear and tear." As such, they may require no further action. Some forms of single-occurrence damage affect key aspects of surface function or appearance and will stimulate follow-up action such as part replacement, repair, alternate material substitution, or even component redesign.

In contrast to single-occurrence surface damage, wear is generally considered to be progressive, involving multiple interactions. The ASTM G40-13b [6] definition of wear reflects this difference:

> *wear*, n.—alteration of a solid surface by progressive loss or progressive displacement of material due to relative motion between that surface and a contacting substance or substances.

Definitions continue to evolve, and the ASTM definition for wear reflects this. The phrase "alteration of a solid surface" replaced the earlier phrase "damage to a solid surface" in the definition because some ASTM committee members argued (wrongly, in the author's opinion) that machining or grinding could wear a surface intentionally and beneficially, yet not be considered "damage" with its negative connotation.

Note also that there is no requirement for material loss in the definition for wear. A process that progressively deforms a surface without the loss of material can also be considered to be wear. Galling or scoring of metals can move material around on a surface without necessarily creating loose particles (see Figure 3.2). Therefore, some would consider galling and scoring to be mechanical surface damage, rather

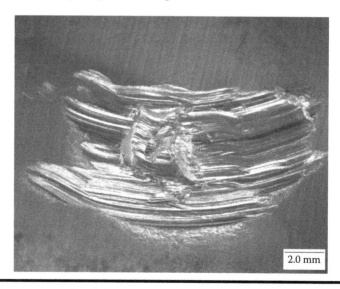

2.0 mm

Figure 3.2   An example of elevated temperature galling damage on stainless steel in which the material is displaced without production of any loose particles. Five arch-shaped strokes produced this result, which indicates that galling can take place in a single event or as progressive damage.

than wear with its "progressive loss" of material. The need for clear terminology is evident when preparing a tribosystem analysis: Are surfaces merely deformed, or is there actually a loss of material? For example, a seal can leak either because it has lost material or because bumps on its surface form during use and prevent it from mating properly.

When applying the various tools and techniques described in Chapter 2 to examine wear surfaces, some forms of wear look similar, and it may be necessary to extend that diagnosis to the examination of debris particles or to the condition of any lubricant, if present. For example, severely worn metals deformed in uni-directional sliding show the grooves that one might associate with abrasion, but the debris particles for adhesion-promoted wear may appear to be flake-like. In contrast, debris from abrasive wear may look like thin, curled cutting chips. The distinguishing characteristics of various forms of wear are presented in the sections that follow.

3.2 Types and Characteristics of Wear

Authors of textbooks, review articles, and handbooks have devised various schemes to categorize types of wear. These are based largely on each author's experience or area of specialization, but as illustrated earlier, they can differ from one article to another—even in the same handbook. In this book, the highest level divisions of mechanical wear will be organized by the type of relative motion. That is, solid bodies that can slide against one another, impact upon each other, or roll over one another. This organizational scheme does not depend on which fundamental mechanisms occur but rather focuses on the motion involved.

The three types of relative motion can further be broken down into subcategories which have corresponding standard definitions. Figure 3.3 shows a system of wear classifications that reflect the types of relative motion approaches. In some cases, the categories reflect special cases. For example, fretting wear is a special case of reciprocating sliding in which the length of the stroke is very small. The terms in the figure reflect both the causal processes (e.g., abrasion, erosion) and the effects of those processes (e.g., abrasive wear, erosive wear).

The types of wear shown in Figure 3.3 correspond to the discussion of various wear forms that follows. Shorthand codes for each of these will be presented to identify the type of wear as part of the Tribosystem Analysis form described in Chapter 5. Hybrid forms of wear in the lower right corner of Figure 3.2 involve both chemical and mechanical effects. As will be illustrated later, their shorthand codes will be used as suffixes for other forms of wear.

In the tribology literature, researchers report the appearance of certain wear processes as if they were basic forms of wear in and of themselves. For example, someone might use the term *spalling wear*, but in the present context, spalling is not a unique form of wear. Rather, spalling is a phenomenon that is associated with contact

Codes for the Categories and Subcategories of Wear

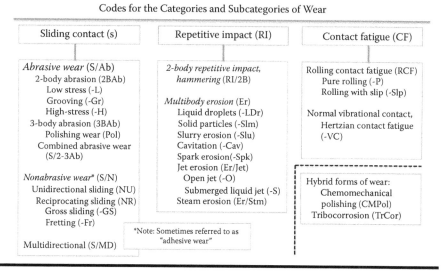

Figure 3.3 Proposed organization of the forms of wear, based on the relative motion between bodies, the number of bodies involved in the interaction, and the material removal process. The codes (in parentheses) are defined later in this chapter.

fatigue rather than being a primary form of wear itself. Similarly, scuffing is a surface damage process that can occur during sliding contact but is not a form of wear. Therefore, the term *scuffing wear* should not be used. Another problem with the term *scuffing* is its connotations in different parts of the world. The European community tends to associate scuffing with plastic deformation and smearing, but the U.S. community sometimes equates scuffing with scoring. The latter connotation is clearly incorrect because to score a surface means to cut a groove in it, and scuffing need not create deep grooves but can actually smooth a surface in some cases.

Another confusing term is *frictional wear*. It is sometimes used in the tribology literature, but friction can occur along with various forms of sliding wear, and therefore, use of the term *frictional wear* is too vague and ambiguous to be useful. As a result, it will not be used in this book.

Finally, the term *adhesive wear* has a historical significance in the literature of tribology. It is often convenient to use this term, especially in conjunction with metals and alloys, to describe surfaces that display significant plastic deformation with pronounced grooves, torn ridges between them, and evidence for the transfer of material to one or both counterfaces. Microscale asperity welding and fracture of such momentary adhesive junctions are part of the interpretation. Rather than identify the type of wear by its features alone, the term carries with it a set of causal mechanisms. Therefore, it is advised to use the term *adhesive wear* sparingly, as is done in this book. Interestingly, ASTM's committee G2 on "Wear and Erosion" named its two subcommittees that deal with sliding wear phenomena as G02.30, on "Abrasive Wear," and G02.40, on "Nonabrasive Wear." The naming of G02.40, which happens to be responsible for standards such as the pin-on-disk, block-on-ring, and reciprocating pin-on-flat tests, makes no prejudgment as to what mechanism (i.e., adhesion) is or is not responsible for the observed wear, only that it is not predominantly caused by 2- or 3-body abrasion.

The distinguishing characteristics of the forms of wear are described in the following subsections. In some cases, ASTM standard definitions are provided, based on ASTM G40-13b [6].

Understanding the evolution of wear in sliding contacts can help to explain frictional behavior [7]. Frictional changes over time can indicate, albeit indirectly, changes in the severity and type of wear that is occurring. For example, as described in more detail elsewhere [8], sliding may begin by wearing through an oxide film to expose the bare metal, which then begins to shear and work-harden. As damage builds up, particles are released to introduce an abrasive component. All these changes can be reflected in the friction records of sliding experiments, and it is likely that a sequence of processes also occurs in engineering systems. Therefore, when identifying the types of wear in a tribosystem, it is helpful to consider that the forms of wear can change with time. Preventing the onset of an earlier form of wear may delay the onset of a more severe form that derives from it.

The prevalence of the various forms or categories of wear is not the same in all areas of technology. In the late 1970s, Eyre [9] reported the relative occurrences of wear in various industries (Table 3.1). The remainder of this chapter describes the characteristics of wear that follow the breakdown shown in Figure 3.2. This convention and the identity codes associated with them will be used in Chapter 5 as shorthand notation in the tribosystem analysis form. The discussion in the balance of this chapter will support the need to know more about the specifics of each problem when selecting palliatives and/or test methods.

Table 3.1 Occurrence of Various Types of Wear in Industry

Wear Type[a]	Occurrence (%)
Abrasive	50
Adhesive	15
Erosive	8
Fretting	8
Chemical	5

Source: Eyre, T.S., *Tribology International*, 11, 91–96, 1978 [9].

[a] These categories were used by Eyre, but they are described in more detail and separated into subcategories in the sections that follow.

3.2.1 Wear Categories, Processes, and Mechanisms

Schematically, Figure 3.4 shows that wear types (e.g., abrasive wear) are composed of processes (e.g., cutting and plowing by hard asperities), which in turn are composed of finer-scale wear mechanisms (e.g., grain boundary deformation, dislocation generation and interaction, atomic scale adhesion). In Figure 3.4, note that the same fundamental mechanisms (e.g., "work-hardening") can play a role in more than one process. This overlap reveals part of the challenge for those engaged in wear modeling (see, for example, Meng and Ludema [10]). However, the intelligent approach to diagnosing the dominant form of wear that is operating in a tribosystem is to use the method of multiple attributes, as described in Section 3.5, at the conclusion of this chapter.

3.2.2 A System for Use in Distinguishing the Common Forms of Wear

The remainder of this chapter describes the distinguishing features of three categories of wear based primarily on their relative motion and characteristic appearance. While it is convenient to separate the various forms of wear in a hierarchical scheme, it is also prudent to recognize that more than one form of wear may occur in a given tribosystem, or even on different places on the same component part. Examples of these phenomena are provided later.

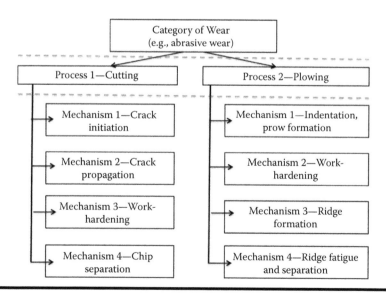

Figure 3.4 An illustration of how wear categories can be broken down into processes and mechanisms.

3.2.2.1 Sliding Contact (Category Code: S)

Sliding contact involves tangential relative motion between two or more bodies. Interactions during sliding contact can involve abrasion, adhesion, and various combinations of the two. A situation can begin with adhesive sliding that later generates wear particles to become a combination of adhesive wear and 3-body abrasive wear. Therefore, different forms of wear can exist at different times in the history or a tribological contact, and the word *sliding* is a high-level descriptor that refers only to the fact that relative motion is occurring and not which processes occur in the interface as a result of that motion.

Abrasive wear (Category Code: S/Ab). Terminology standard ASTM G40-13b [6] defines abrasive wear as follows:

> *abrasive wear,* n.—wear due to hard particles or hard protuberances forced against and moving along a solid surface.

Abrasive wear, sometimes referred to as "grooving wear" or "cutting wear," is further differentiated by the number of solid bodies that are involved. Two-body abrasion, for example, involves a hard rough surface like sandpaper sliding on a softer surface so as to produce cutting chip-like debris. Three-body abrasion involves many fine particles, like sand or polishing compound, that roll around between two other bodies, digging into them and eventually producing material removal. Formally, the abrasive species is called an *abradant.* Also from ASTM G40-13b [6]:

> *abradant,* n.—a material that is producing, or has produced, abrasive wear.

An example of the silica sand abradant that is used in dry-sand/rubber wheel wear testing (ASTM G65 [11]) is shown in Figure 3.5.

Considering its prevalence in mining operations, oil exploration, and agriculture, and as estimated in Table 3.1, abrasive wear is likely the most costly form of wear in terms of total material wastage.

Two-body abrasive wear (Category Code: S/2BAb). The most common example of this form of wear is the use of sandpaper. The abradant is fixed on one body and the second body moves relative to it. The result is a series of parallel scratches (called "striations" by the criminal forensics community) or grooves.

There are varying degrees of 2-body abrasive wear. These have been qualitatively described as mild or low-stress abrasion (L), deeper grooving (Gr), and high-stress abrasion (H). An even more severe form, called "gouging abrasion," is discussed by Budinski [12]. An example of mild 2-body abrasion is shown in Figure 3.6 [13]. Note the fine parallel scratches and evidence of plastic tearing of the ridges at the edges of some of the scratches. One proposed mechanism for wear is that continued abrasion by offset grains bends the thin edges of the grooves back and forth, eventually causing them to fatigue and break off at the microscale.

Figure 3.5   These rounded particles of graded AFS 50-70 silica sand, used in the ASTM G65 dry-sand rubber wheel test, are considered to be an abradant. This test is becoming more difficult to perform due to a reduction in the sand supply (now used for fracking). The shape, size, and frangibility of abradants can affect the severity of 3-body abrasive wear.

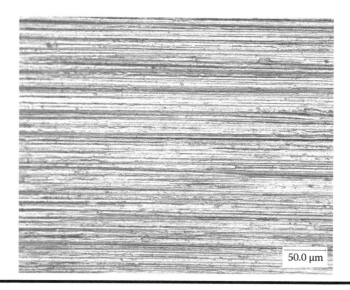

Figure 3.6   An example of mild abrasion of a titanium alloy (Ti-6Al-4V) specimen from a continuous belt, low-stress abrasion test. (ASTM G174-04, *ASTM Annual Book of Standards*, Vol. 03.02, ASTM International, W. Conshohocken, PA, 2014 [13].)

Gouging abrasion results in deep cutting, observable to the naked eye. Such damage can include cracking, tearing, and/or chipping of the surrounding material. A jaw crusher test has been developed for gouging abrasion in the mining industry [14]. It consists of two pairs of opposing, inclined plates. One pair contains a reference material such as cast manganese steel, and wear is expressed as the ratio of weight loss of the test plate to the reference plate. High-stress abrasion is characterized by conditions so severe that the abradants themselves are fractured in the process and can form third bodies. Therefore, the size distribution of the abradant would change and indicate that this form of wear was occurring.

Fundamental studies of single-point scratches on pure metals indicate that the level of damage increases as the contact pressure on the abradant increases. For example, Figure 3.7 shows damage on three pure metals scratched left to right with a rounded diamond tip having a 75-μm radius [15]. The Fe is body-centered cubic,

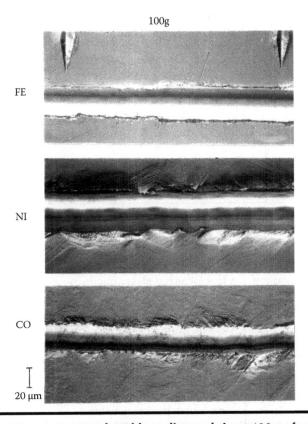

Figure 3.7 Deep grooves produced by a diamond tip at 100 gr-f on three pure metals. The mode of deformation affects the lateral extent and nature of damage from grain to grain. (From Blau, P.J., *Microstructural Sci.*, 12, 293, 1985 [15].)

Ni is face-centered cubic, and Co is hexagonal. The narrower the width of the scratch, the harder the surface. Note how the scratch width on Fe changes across a grain boundary in the center of the scratch. The effects of crystallographic orientation on scratch resistance are indicated.

As one might expect, the type and severity of abrasion damage are material dependent. This point is illustrated by a series of experiments on the types of damage produced on two polycrystalline ceramics (Al_2O_3, SiC) and two Al_2O_3-based ceramic composites with 10 and 20 vol% SiC particle contents. The materials were subjected to abrasion by a diamond stylus with a 200 μm tip radius at different normal forces (P.J. Blau, 1987–1990, data collected for scratch tests on alumina matrix matrix/SiC composites at Oak Ridge National Laboratory, Oak Ridge, Tennessee). The types of damage are listed in Table 3.2.

Figure 3.8 compares the scratch features observed on the Al_2O_3 and SiC single phase specimens of the matrix and additive materials, and Figure 3.9 compares them on the two different composites at the same loads. This is an example of how a list of characteristic damage features can be tabulated in order to depict both the severity of wear and compare differences in wear response of different materials using the same damage criteria and similar contact conditions.

While single-point scratch tests like those described above are interesting from a fundamental standpoint, single-point tests do not simulate the kind of damage produced on a material by multiple interactions. Between single-point and full

Table 3.2 Types and Severity of Scratch Damage on Ceramic Composites (Al_2O_3/SiC)

Level	Qualitative Rank	Description
1	Mild	Smooth, nearly featureless groove
2	Mild	Fine parallel striae in the groove lying in the scratching direction
3	Intermediate	A few localized glassy fractures
4	Intermediate	Fine-scale chipping along the scratch edges
5	Intermediate	Arc-like, periodic tensile cracks within the grooves
6	Severe	Hairline cracks extending from the groove edges into the matrix
7	Severe	Periodic spalls, whose length is less than about one-fourth the crack width
8	Severe	Gross fractures extending across the groove and into the matrix around it, including large edge chips

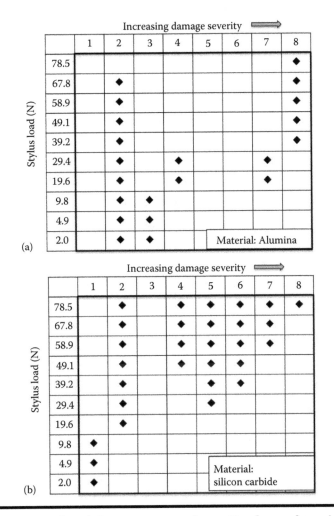

Figure 3.8 Hierarchy of damage types and severity observed on single-phase polycrystalline ceramics subjected to scratch tests at different loads: (a) alumina, a matrix material; (b) silicon carbide as a pressed and sintered tile.

abrasive contact, there is a limited body of research on multiple stroke or overlapping single-point tests (see, for example, Gee [16]). Most practical cases of abrasive wear involve numerous overlapping contacts.

An example of how a bearing steel can be torn during 2BAb, produced during ball-on-flat tests using a 240-grit SiC grinding paper as the opposing surface, is presented in Figure 3.10. Note the relatively continuous grooves varying in width and depth.

Depending on the material being abraded, debris from 2-body abrasive wear processes tends to display the appearance of microscale machining chips, sometimes

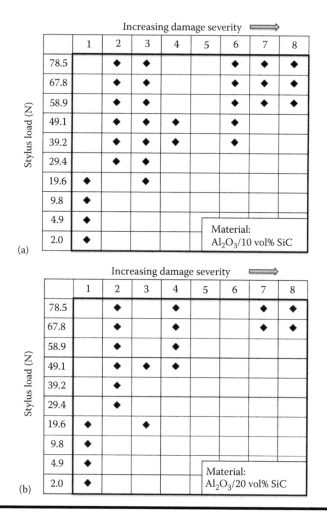

Figure 3.9 **Hierarchy of damage types observed on composites of Al₂O₃ with SiC particles subjected to scratch tests at different loads: (a) composite with 10 vol% SiC particles; (b) composite with 20 vol% SiC particles.**

curled up and sometimes with feathered edges. Examples of cutting chips produced by the 2BAbr of steel on abrasive paper are shown in Figure 3.11. Debris appearance is one way to identify abrasive wear from samples of lubricating oil during oil analysis (see Chapter 2).

Even relatively mild abrasion takes its toll on materials. For example, only a few milligrams of material lost from a coin can make its value more difficult to read, and while the size, material, and shape can help to distinguish it, old worn coins lose value as a collectable. (Note: There is an interesting article on the wear

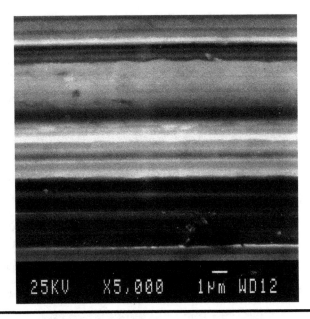

Figure 3.10 **Scanning electron microscope image of 2-body wear features on a specimen of hardened 52100 steel rubbed repeatedly, stroke by stroke, against SiC paper. (From Blau, P.J., Whitenton, E.P., and Shapiro, A., *Wear*, 124, 1–20, 1988 [17].)**

of coins [largely by abrasive wear] published by Spurr [18]. By weighing coins of various years to a precision of 0.2 mg, Spurr calculated the volume lost for pennies versus the year of coinage for 900 specimens between the years 1860 and 1965. Spurr noted a nonlinear relationship and a trend that showed a decrease in the wear rate as the hardness of the penny improved through the use of brass and then cupronickel.)

Three-body abrasive wear (Category Code: 3BAb). Three-body abrasive wear is common in mining, agriculture, and industrial processes in which loose hard particles are handled. Depending on the surroundings, their motions, and the surfaces that entrap these particles, they can tumble, become embedded, or pass through with little harm done. Figure 3.5 shows the kind of sand particles that are used in a popular 3-body abrasion test (ASTM G65 [11]). The sizes, shapes, angularity, frangibility, and composition of these abradants affect the severity of wear. Particles that are too large may not be drawn into (entrained) the bearing interface, and particles that too small may pass through without causing harm. Particles of a certain size range may be drawn into a given bearing configuration and produce significant 3-body abrasive wear.

When hard particles are not fixed as in 2BAb, they can tumble, embed, and dig grooves into surfaces to produce a variety of features and impression shapes.

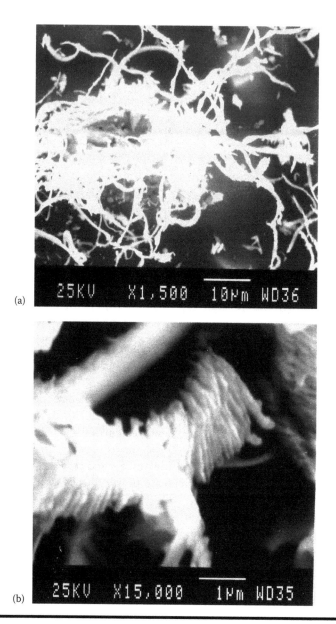

(a)

(b)

Figure 3.11 Cutting chips from sliding AISI 52100 steel against SiC abrasive paper (a) entangled thin cutting chips extruded from the contact, and (b) serrated chips caught between abrasive grains in the SiC paper. (From Blau, P.J., Whitenton, E.P., and Shapiro, A., *Wear*, 124, 1–20, 1988 [17].)

This was illustrated by Samuels by rolling a plastic cube on modeling clay (see Figure 3.12) [19]. Therefore, 3BAb produces fewer scratches that lie in the direction of relative motion (for example, see Figure 3.6), but rather a more randomized mixture of indentations, tears, and grooves. At a fine scale, as in polishing, this gives a reflective surface a smooth, nondirectional appearance. If the surface is rougher, 3BAb surfaces are more dull and matte-appearing.

Polishing wear (Category Code: S/Ab/Pol). Polishing wear produced by loose interfacial particles is a special case of 3BAb. In this case, the scratches tend to be multidirectional and may be so fine as to be invisible to the naked eye or even in a microscope at low to moderate (a few ×100) magnification. Also, debris might be difficult to find and the roughness of the worn part may be smoother than the starting surface finish. Polishing can involve both mechanical and chemical contributions (tribochemical effects and tribocorrosion), but in the present category designation, we shall consider its classification to be based on multidirectional motion and mechanical mechanisms of material removal. One of the best reviews of the mechanical aspects of polishing wear can be found in the article by Samuels [20]. An article by Ryason et al. [21] describes the effects of polishing wear of bearing steels by soot of various compositions and particle sizes.

The term *polishing wear* has been associated with the wear of lubricated gear teeth that appear to be mirror finished after running. Polishing wear can be a tribochemical process that is facilitated by abrasive particles in a lubricant that remove oxide films while at the same time chemically attacking the surfaces. Therefore, a two-pronged approach can be used to control this type of tribochemical process. One involves filtering the lubricant to remove abradants and the other is to adjust the lubricant chemistry.

Polishing wear can range from chemical polishing with little or no mechanical contribution to a more grinding form of polishing that involves 2-body or 3-body

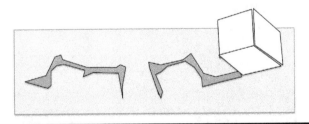

Figure 3.12 Indentation shapes produced on a soft clay substrate by tumbling a plastic cube on it. (Recreated from an image in Samuels, L.E., *Metallographic Polishing by Mechanical Methods*, 3rd ed., ASM International, Materials Park, OH, 1982 [19].)

abrasion. In practice, a contact surface can appear polished, but some investigators might call it "scuffed." Consider the example of an airport luggage carousel in which metallic blades slide over one another around the corners (Figure 3.13). Note the polished/scuffed triangular areas on the satin-finished AISI 304 stainless steel 14-gauge blades where overlap occurs during cornering. The incline of such equipment is typically ~20 degrees. In addition to dents, there are some larger scratches on the polished areas due to 3-body abrasion from hard particles caught between the blades. An airport luggage carousel is an example of the occurrence of both surface damage (incidental scratches and dents) and progressive wear.

The development of silicon wafer manufacturing has prompted intensive study of the mechanisms of polishing. In this case, the polishing is intentional and carefully controlled, not a wear process to be reduced or avoided. Chemomechanical polishing (also called chemomechanical planarization, or CMP) has been the

Figure 3.13 Airport luggage Carousel 2 at Knoxville, Tennessee's, McGhee-Tyson Airport showing a type of scuffing contact that polishes the mating surfaces (wedge-shaped shiny features). Therefore, scuffing may not always indicate roughening.

subject of proprietary as well as open-literature research. A review of the process and discussion of proposed mechanisms may be found in an article by Xin et al. [22], and books on CMP, such as Doi et al. [23], describe industrial approaches to controlling this form of wear in manufacturing.

Combined 2-body and 3-body abrasive wear (Category Code: S/2-3Ab). In some tribosystems, it is difficult to precisely differentiate between surfaces subjected to 2BAb and 3BAb because both types of wear can occur at the same time. A plow cutting through the soil can be abraded by loose, tumbling stones as well as those that do not move so easily aside and that can cut deep grooves (Figure 3.14). In another example, hard particles on one surface can cut debris, which then rolls around and further wears the interface. This type of transition and conjoint action complicates tribosystem analysis and requires more careful study to sort out. Perhaps one of the forms is dominant and therefore should be the focus of material selection. The 2BAb–3BAb situation also raises the question of whether abrasive debris from an external source initiated the wear or whether the abrasive particles doing the major damage were formed in the contact from work-hardened material from one or both sliding bodies. Additional tools, like surface chemical analysis, lubricant analysis, and electron microscopy, may be required to answer these questions.

Nonabrasive wear (Category Code: S/N). Historically, severe forms of metal-on-metal sliding wear damage have been called "adhesive wear" because early researchers advanced the concept that the mating surfaces formed adhesive bonds like tiny solid-state welds and that those bonds or the underlying material (if weaker than the bonds) yielded and failed as material was removed. Details of that approach

Figure 3.14 Disc plows, such as this 1830s vintage model on display at the Biltmore Estate in North Carolina, experience both 2BAb and 3BAb, as well as impact wear (with rocks) as it is pulled through the soil. While the number of surface engineering options for such wear-critical components has greatly increased since that time, material choice is still largely governed by cost.

may be widely found in the tribology literature of the 1950s and 1960s (e.g., the classic book by Bowden and Tabor [24]). Whether actual chemical bonds form or whether rough features simply interlock and deform without chemical bonding taking place, the term *adhesive wear* is still commonly used to describe sliding processes other than those associated with abrasive wear. That being said, there is some ambiguity because work-hardened asperities and debris that may form during the metallic adhesive wear process may produce grooves similar to those observed on abrasive wear surfaces. For example, Figure 3.15 shows a worn latch bolt in which the brass plating has been worn through to the underlying steel. The shiny

(a)

(b)

Figure 3.15   Wear of similar plated brass latch bolts used on exterior (a) and interior (b) doors. On the exterior door, several types of damage, including fine scratches, galling (displaced and raised material), and adhesive transfer have occurred. The brass plating has worn through, as can be seen at the end and sides of the 8-mm-wide wear scar. In the lower bolt, also in use for 11 years, but on a lighter duty pantry door, wear-through occurs but is less severe, with only minor abrasion and no macroscopic galling or gross material displacement.

appearance may cause some investigators to call this "scuffing," but under high magnification, fine scratches may lead to the determination that it is 2-body abrasion. At the upper right end of the stroke (near the center of the image), the plastic deformation of material above the plane of the unworn contact could be termed *microgalling*. Close examination of plateaus of material near the center of the contact area, however, identifies this damage primarily as adhesive wear. The occurrence of multiple types of features like the foregoing example causes problems when documenting wear in a tribosystem analysis.

Severe adhesive wear is recognizable by significant plastic deformation, obvious tearing of surface features, plowing grooves, and evidence of material transfer. An enlarged optical photomicrograph of adhesive wear, produced by a silicon nitride ball sliding against stainless steel, is shown in Figure 3.16. The rough topography is typical. Debris particles, some of which are shown attached to the surface in Figure 3.16, typical of metal-on-metal adhesive wear, tend to be shiny and flakelike. Figure 3.17 shows a fractured wear debris flake from a Cu-Al alloy sliding dry against bearing steel in which the deformation of the deformed layer from which

Figure 3.16 An example of type S/NU wear showing multiple attributes. Nonlubricated sliding direction was from the bottom to top. (Field of view, ~410 μm wide.)

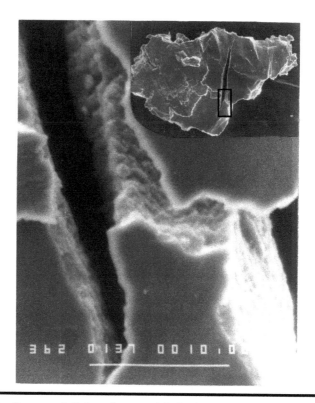

Figure 3.17 Fractured flake of wear debris from unidirectional sliding wear, generated in a pin-on-disk test of steel on a Cu-7.5 wt% Al alloy. The inset on the upper right shows the entire flake, and the main image shows the cellular structure of the fracture face of the deformed layer from which the flake formed. (From Blau, P.J., *Interrelationships among Wear, Friction, and Microstructure in the Unlubricated Sliding of Copper and Several Single-Phase Binary Copper Alloys,* PhD Dissertation, The Ohio State University, Columbus, OH, 1979 [25].)

the flake originated can be seen in the fracture as a brick-wall like structure of dislocation cells.

Unidirectional sliding (Category Code: NU*).* This type of relative motion has had a long history of study, and in fact, early models of wear, like those of Archard [26] and Holm [27], who worked in the early and mid-1900s, concentrated on NU processes, especially in ductile metals. Nonabrasive sliding wear occurs commonly in soft metal bearings and bushings. The term *unidirectional* is taken broadly to apply not only to linear motion like a block on a flat surface but also to rotating components like shafts that continue to turn in the same direction. Hence, a shaft turning in a plain bearing and a block pressed on a rotating ring can be considered unidirectional even though at least one of the mating parts rotates.

Use of the term *nonabrasive wear* or *adhesive wear* or *metallic wear* is somewhat ambiguous because when one looks closely at the mechanisms that produce the wear process, there are also aspects of abrasion. NU does not require abrasion, but abrasion can be involved. Fundamental interpretation of metallic contact has involved breaking through oxides to form adhesive bonds between mating surfaces. These localized junctions grow under increasing contact pressure, and if shear is applied as in sliding with friction, then the junctions can rupture either at the interface between the metals or below the surface of the weaker of the two partners.

The continued formation and rupture of junctions produce wear debris, and the shears applied to the surfaces can also locally work-harden the sliding metals. Therein lies some of the ambiguity when trying to define "nonabrasive sliding wear." The roughening of a work-hardened metal can cause it to abrade a softer counterface, and the presence of wear debris can produce 2BAb or 3BAb, respectively. Thus, a sliding component can begin its wear life with a smeared, ductile appearance in which the deformation and rupture of junctions predominate and then later on take on an abraded appearance as the process and its debris generation continue. Figure 3.16, mentioned earlier, shows an example of severe metallic wear from an experiment on nonlubricated sliding. Also, Section 3.4, at the end of this chapter, discusses how changes in wear type can occur. Such transitions can negate the presumption of a constant steady-state wear rate.

The pristine wear debris associated with NU of metals and alloys, if not caught in the interface and crushed into a fine powder or agglomerated mass, takes on a flake-like appearance. A popular theory called "the delamination theory of wear" was introduced by Suh and his group in the early 1970s to explain the debris' origins [28]. Figure 3.18 from the author's work [25,29] exemplifies the features of flakes associated with NU. Wear diagnosis using oil sampling and ferrography looks for such flakes to identify adhesive or metallic wear.

However popular a new wear theory (model) might be, there may be observations that at first look appear to conform to it but, after closer examination, may be due to other basic processes than that specific theory requires. In the case of delamination, wear particles may form by the shear of third-body layers to form thin films, which eventually flake off. An example is shown in Figure 3.19 [30]. This rear drum brake shoe from a commercial trailer was subjected to 4 years of service. During this time, it acquired a delamination-like morphology that comes from crushing and shearing the near-surface friction material into a fine-structured third-body layer, which eventually detaches to leave a general appearance similar to what might be observed for Suh's "delamination theory of wear" [28], but different in detail.

Reciprocating sliding (Category Code: NR). Reciprocating sliding occurs when the sense of the direction of tangential relative motion in the interface changes during use. This chafing type of motion that produces linear features on the wear surface is characteristic of components in piston engines, certain types of reciprocating pumps, slider-crank mechanisms, and ordinary door hinges.

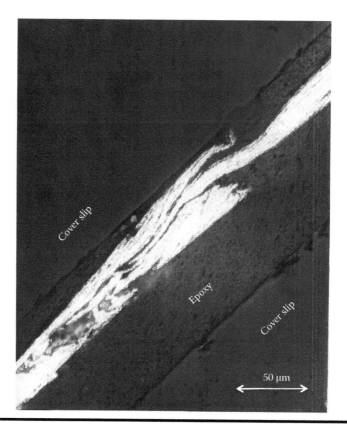

Figure 3.18 A cross-section of a sliding wear particle of Cu-70 wt% Zn showing the heavily deformed "marble-cake" microstructure associated with the delamination of a flake from the mechanically mixed layer. (From Blau, P.J., *Wear*, 66, 257–258, 1981 [29].)

The effect of this motion is to apply reversed traction to the surface, to impart a wear texture, and to redistribute any debris that might form either to the side or to the ends of the stroke. By virtue of geometry, the speed of relative motion in the reciprocating sliding tends not to be constant but varies in accordance to the mechanical design. In most cases, sliding speed tends to decrease toward the ends of the stroke and increase at the center. Depending on the specific case, a plot of the contact location versus time in reciprocating sliding can range from a periodic, quasi-sinusoidal appearance (like a slider and crank at steady-state) to an erratic shape, like a door that opens and closes from time to time. Stepper motor-driven assemblies can take on more of a square wave shape.

When lubricants are present under variable speed conditions, their ability to form thick, protective entraining films varies with the load, viscosity, and relative sliding speed, among other considerations [31]. Therefore, the wear observed on reciprocating sliding devices can vary in severity along the length of the stroke

Figure 3.19 Worn surface of a drum brake friction pad from a heavy-duty truck trailer showing multistage delamination. First, a tribolayer was formed, then it was delaminated. Evidence for abrasion appears when the smeared tribolayers are removed. For scale, the mounting bolt holes are approximately 13 mm in diameter. (From Blau, P.J., *J. Mater. Eng. Perform.*, 12, 56–60, 2003 [30].)

according to the lubricant film thickness. In cylinder liners that are used in internal combustion engines, the wear tends to be most pronounced at the top end of the piston ring stroke in the combustion chamber because the oil film breaks down, the shearing direction changes, and the temperature is highest. As a wear groove forms on the cylinder, the ring tends to seal less well and the efficiency of the engine begins to degrade. The tribology of internal combustion engines, including specific areas in which NR occurs, is reviewed in a large number of publications, for example Taylor [32], Kapoor et al. [33], and Davis [34].

Gross reciprocating sliding (Category Code: NR-GS). If the stroke length in reciprocating sliding is at least twice as long as the length of the sliding contact in the direction of relative motion, then one body is in continual contact, but if some of the mating body is sometimes out of contact, then gross reciprocating sliding can occur. If there is some overlap, then some portions of the wear scar may never be fully exposed, and if the wear area is largely constrained from slipping, except perhaps at the edges, then much of the surface is said to be in a state of "stick" and only the edges are in "slip." The boundary between these areas is commonly referred to as a "mixed zone." Wear particle motion and the ability of such debris to escape from reciprocating contacts can affect the appearance of reciprocating sliding wear features.

In applications like compression piston rings in the cylinders of internal combustion engines, most of the wear is at the hot end, where the lubricant film collapses at zero relative velocity and the high combustion temperature reduces the lubricant effectiveness. Conversely, scuffing can occur on pistons at the lower end, a "skirt" area during start-up when the lubricant is cold and an effective film has not yet formed. This is another case where wear can vary in degree along a reciprocating stroke.

Similar considerations apply to the special case of short-stroke reciprocating sliding known as fretting, a subject that has important applications in mechanical fastening, electrical contacts, turbine blading, when transporting rolling bearings, and a host of others.

Fretting (Category Code: NR-Fr*).* Fretting is a special case of low-amplitude, reciprocating sliding. In truth, there is some ambiguity in listing it here because portions of a fretting wear scar can exhibit adhesive wear features and other areas can exhibit abrasive features due to the action of fretting debris particles. Wear scars from fretting can appear rough or pitted, and some cases look like "halos" depending on the fraction of the contact that is stuck or slipping during relative motion. Fretting wear is listed here in the hierarchy in Figure 3.3 on the basis of its relative motion: low-amplitude sliding. Likewise, the definition of what constitutes true fretting tends to be somewhat cloudy if the definition involves, for example, the specific amplitude of slip or whether there is a prevalence of stick or slip in the contact.

During the 1970s and 1980s, the well-known fretting researcher, R.B. Waterhouse, from the University of Nottingham, United Kingdom, attempted to define a critical amplitude between fretting and reciprocating sliding wear [35]. Chen and Zhou, among others, also addressed the defining transition between these two forms of wear [36]. This amplitude ranges from roughly 25 to 300 μm.

There is a related damage phenomenon called *false brinelling* that is sometimes confused with fretting. There are some differences [37]. False brinelling results from the minute rocking of a concentrated contact that produces a feature that can look like a hardness impression. There are no sliding or abrasion-like striae as might be observed in the case of fretting contact. True brinelling, by contrast, is the action of a high-load indentation or impact to produce a depression like a hardness indentation. Such phenomena as false and true brinelling are commonly associated with damage to the races in rolling element bearings.

In fact, fretting can occur over a range of sliding amplitudes, and in dry fretting situations, the loss of material is influenced by the action of third bodies (debris) that form in the contact and work their way out. The classical Mindlin mechanical analysis [38] and further work, like that in France (French work on fretting zones mixed, slip, stick, etc.), have led to the realization that in an oscillating contact, there may be zones in which there is no relative motion ("stick"), zones in which there is complete sliding ("gross slip"), and transition zones between them ("mixed"). Details of these zones and their features may be found in the tribology literature, but the point here is that some tribological contacts may have some combination of fretting damage, gross sliding wear damage, and stick.

Below a critical amplitude of relative motion, the fretting wear rate is negligible. As the amplitude increases, fretting wear rate increases until the amplitude reaches a point at which reciprocating sliding wear damage becomes dominant. Roughly, the amplitude for fretting wear is from 50 to 300 μm. Measuring the displacement in fretting tests can be challenging due to the elastic compliance of fixtures or specimen holders. They must be extremely stiff to get accurate measurements of micrometer-sized fretting displacements in the sliding interface. It is possible that an interface may be stuck together without relative motion even though the elastic structure around the stuck contact point oscillates.

In terms of type of relative motion, Zhu and Zhou [39] defined four situations (see Figure 3.20): (1) linear oscillatory motion "tangential fretting," (2) radial expansion and contraction, (3) pivoting motion, "torsional fretting," and (4) rotational fretting. The first type of relative motion involves oscillation back and forth in a straight line, the second involves an expansion and contraction of the contact area (such as pushing vertically up and down on an elastic-plastic circular contact), the third involves pivoting or torsional motion above a center of rotation normal to the mating surface, and the fourth involves oscillatory rotation in which the axis of rotation is parallel to the plane of the mating surface. These motions affect the slip versus stick conditions in the contact zone and the movements of debris within and to the edge of the contact.

The stresses on the ball on flat (sphere-on-plane) geometrical arrangement are widely studied, and most analyses are based on the work of Mindlin [38]. Frictional considerations aside, a first-order understanding of an oscillating ball-on-plane, such as that shown in Figure 3.21, can be found simply by comparing the change in elastic (Hertzian) contact diameter ($\Delta 2a$ in the figure) with a varying applied force, ΔP (see also the related discussion of contact fatigue later in this chapter).

The Hertz equation for the elastic contact radius (a) for a sphere of diameter D on a plane of the same material under normal force P is [40]

$$a = 0.721\sqrt[3]{PDC_0}, \tag{3.1}$$

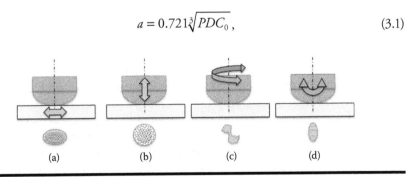

(a)   (b)   (c)   (d)

Figure 3.20 **Four kinds of fretting, based on the relative motion: (a) tangential or linear oscillation; (b) radial or expanding and contracting contact; (c) torsional or twisting contact; and (d) rotational or tilting contact. (Based on Zhu, M.H. and Zhou, Z.R., *Tribol. Int.*, 44, 1378–1388, 2011 [39].)**

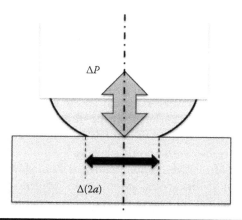

Figure 3.21 A normal oscillatory load on a nonconformal contact produces "Hertzian contact fatigue" and radial fretting as seen in Figure 3.20b.

where the composite modulus C_0 is

$$C_0 = 2\left[\frac{(1-\upsilon^2)}{E}\right]$$ (3.2)

and ν = Poisson's ratio and E = elastic modulus of both the ball and the plane. Inserting values for self-mated Type 52100 bearing steel ball (E = 210 GPa, ν = 0.3) with sphere a diameter of either 10 or 20 mm and oscillating loads, against a flat plane, the radial distance of expansion is plotted in Figure 3.22. For the case of a normal load oscillation of 50 N, the maximum contact stress (Sc) for the 10-mm ball is about 60% larger than that for the 20-mm ball even though the radial expansion [$(\Delta 2a)/2 = \Delta r$] is about 20 μm smaller in amplitude. In such cases, there can be surface damage from sliding in the annulus and also fatigue damage below the surface in the volume of material subjected to changing contact stress. Therefore, fretting plus contact stress can occur simultaneously.

In dry fretting of ferrous metals, such as in mechanical joints, fretting is characterized by the presence of fine red-brown oxidized debris (which has sometimes been called "red mud"). By contrast, aluminum fretting debris looks black. The presence of fine, powdery debris is one indicator of fretting wear, and clues include the loosening of joints and a pitted appearance of the buried interfaces if the joint is disassembled. If the area of contact is large, debris may work itself down rather than escape laterally, producing tapered carrot root-like features in polished cross-sections. The message is that understanding the role of third bodies is key in diagnosing fretting wear.

Multidirectional sliding (Category Code: S/MD/+wear type*).* Multidirectional relative motion can occur with either abrasive or nonabrasive wear, so the "M" is used

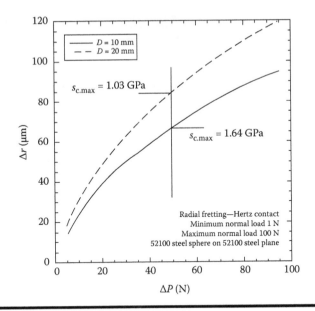

Figure 3.22 A change in the elastic contact radius corresponding to various changes in the normal force for two sphere sizes. The minimum load in each case was 5 N, but the maximum load varied by up to 100 N.

as a suffix for the wear types described previously. The designation S/MD/3BAb, for example, would refer to multidirectional sliding wear dominated by 3-body abrasion. An important example of multidirectional motion is movements by a ball-in-socket joint in a human hip. Since various oscillations and rotations can occur during walking, the direction of relative motion changes frequently during each step. Early experiments and even standards for wear testing of polymers for such applications which tried to use simple oscillating motion failed to produce similar wear as observed clinically in ultra-high-molecular-weight polyethylene. It turns out that the cross-path motion is needed for accurate simulation and material selection in such applications. Another important example of multidirectional wear is polishing or sanding using an orbital type of machine.

3.2.2.2 Repetitive Impact (Category Code: RI)

Two-body repetitive impact (Category Code: RI/2B or RI/3B). Repetitive impact wear involves a normal component of relative motion by one or more bodies against another. Examples of 2-body repetitive impact (RI/2B) are hammering, part stamping, the impact of vibrating of heat exchanger tubes in their holes in a tube sheet, or mechanical printing. The anvil and the heavy hammers shown in Figure 3.23 experience RI/2B, in which neither of the bodies actually loses mass but rather deforms to the point whether they no longer provide useful working

Figure 3.23 Hammers and anvil from the blacksmith shop at the Biltmore Estate, Asheville, North Carolina.

surfaces. Note that alteration of form (displacement of material) is included in the ASTM definition for *wear*. No progressive mass loss is required for this kind of wear to occur.

The effects, cost, and economics of repetitive impact are of interest to those involved in forging and stamping industries, those who print with mechanical type, and those who manufacture coins and currency. The history of coining is an interesting one and reflects the development of wear-resistant steels as well as the coins themselves. The relative hardness and deformational characteristics of these two bodies, the use of heat and lubricants in forming, and other aspects of coin making constitute a body of knowledge on the control and minimization of RI/2B. One of the only books ever written on impact wear was by Engle [41,42], a research physicist at IBM Corporation whose interest focused on the area of printer-type wear.

The properties of different kinds of materials affect their response to 2-body impact wear and obviously lead to different kinds of surface features. Ductile metals, such as bronzes, mild steels, aluminum, and titanium, tend to smear into shiny, specular, and/or dimpled appearances. More brittle materials like high-hardness steels and ceramics display cracking, spalling, and pitting, whose size scale and extent of damage depend on the magnitude of the energy dissipated. The amount of energy dissipated in an impact is a function of the change in kinetic energy of the impacting body over the corresponding deceleration distance (depth of penetration) and has been more fully described in Engle [41,42]. The references by Engle also contain a discussion of compound impact with slip.

In some tribosystems, like mining operations and deep drilling, impact between two bodies may occur in combination with 3-body abrasion (e.g., rocks passing between the jaws of a crusher). In this case, called RI/3B, the surface appearance can take on a complicated series of indentations with superimposed straight or curved grooves. Figure 3.24a,b,c show some examples of this combined wear mode created using a laboratory apparatus with rotating H13 steel vanes rubbing in a cup filled with a loose mixture of alumina and silicon carbide grains to simulate rock. In Figure 3.24a, there was only impact (RI/3B); but in Figure 3.24b, there was only abrasion (3BAb); and in Figure 3.24c, impact plus abrasion produced by periodically raising and dropping the specimen holder onto the turning cup of abradant (RI/3B + 3BAb).

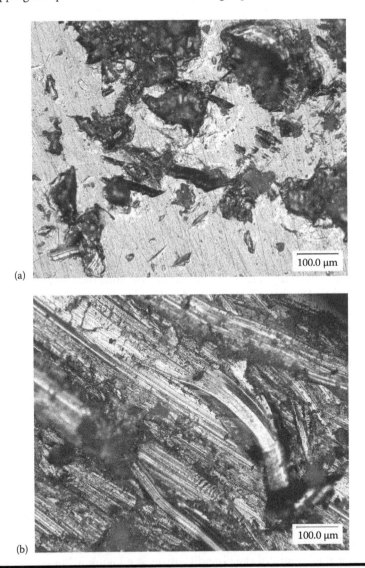

(a)

(b)

Figure 3.24 Wear surfaces of H13 steel test specimens impacting on a loose mixture of angular hard particles in a cup. (a) 100 impacts only, (b) 3BAb without any impacts. *(Continued)*

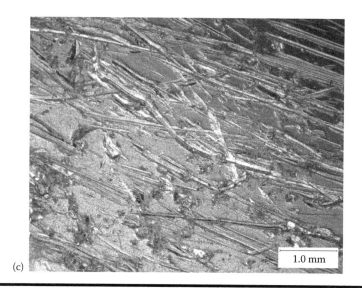

(c)

1.0 mm

Figure 3.24 (Continued) Wear surfaces of H13 steel test specimens impacting on a loose mixture of angular hard particles in a cup. (c) Both RI and 3BAb.

In this study, and counterintuitively, the periodic opening and closing of the gap during impact plus abrasion (in the case of Figure 3.24c) allowed less particle trapping and gouging than in the case of Figure 3.24b. Therefore, the combined wear mode actually produced less material loss than did abrasion alone for the same total sliding distance of the vanes against the opposing flat surface that was covered by grit.

Erosive wear (multibody impacts) (Category Code: Er + descriptor). The process of impact by multiple bodies that are usually small relative to the wearing surface bodies is usually called "erosion" and results in erosive wear. Erosion, as a surface degradation process, probably spans the widest range of size scales compared to any other form of wear. It ranges from the atomic scales of ion-bombardment to the removal of cubic kilometers of material from mountains and oceanic islands. In fact, it can be said that the surfaces of entire moons or planets have been shaped by erosion in some form. From an engineering point of view, and typical of tribosystems of interest in this book, erosive wear can be grouped in terms of what form of erodant causes the erosion. ASTM G40-13b [6] defines an erodant as follows:

> *erodant*, n.—a material that is producing, or has produced, erosive wear.

While the ASTM definition concerns past and present, the sand particles that are found at the bottom of the Grand Canyon (see Figure 3.25), the ore particles impacting the interior corner of a pipe elbow in a processing plant, and a bag of grits about to be used for erosion testing are all erodants, whether they act in the past, present, or future.

Figure 3.25 Mineral sand grains found at Phantom Ranch at the bottom of the Grand Canyon, Arizona. Weathering affects the composition and properties of the mixture. (Collected by D.E. Blau.)

The appearance of eroded surfaces in a wear diagnosis depends not only on the type of material involved but also on the elasticity, deformation, brittleness, and shape of the erodant. The angle of impingement, the particle velocity, the propensity for particles to become embedded into the surface, and the screening effects of rebounding particles can affect the erosion rate as well [43]. Thanks to the intriguing physics of the impact of particles on surfaces, the number of published wear models for erosion continues to grow year by year. Unfortunately, as has been pointed out by Meng and Ludema [10], the large number of erosive wear models in the literature lacks consistency in how the various input parameters come into play. As a result, there are a bewildering number of approaches as to how erosion is modeled. One can confidently say, however, that the rate of erosion depends, at minimum, on the particle velocity, angle of impingement, flux of erodant particles, cumulative exposure, and properties of the erodant and substrate materials under those conditions.

In nature, as in man-made heterogeneous composite materials like metal matrix composites, examples of differential erosive wear are plentiful. More erosion-resistant material remains while the softer and less resistant material is worn away. Figure 3.26 shows the effects of differential erosion by rain, chemical dissolution, and to a lesser extent, hard particle impingement. Interestingly, and despite popular belief, chemical attack and freeze–thaw cycles play a more important role in the degradation of geological formations than does mass loss from solid particle impingement [44].

Figure 3.26 Terraced contours of rock layers in the Grand Canyon (Arizona), show the effects of differential resistance to the conjoint effects of erosion, thermal cycling, and chemical attack. The layers at the top are sedimentary deposits but the materials at the bottom are metamorphic or igneous rocks.

Erosion by liquid droplets (Category Code: Er/LDr*).* Rain erosion is a problem for aircraft windshields and the leading edges of helicopter blades. Depending on the angle of incidence, the damage typically takes on a micropitted appearance. In composite materials, it can also look like grit-blasted surface in which the successive layers of a coating have been torn off. ASTM Test Method G73-10 [45] was developed to measure erosion damage from rain and has been used specifically on aircraft leading edges (wings and helicopter blades).

Erosion by solid impingement (Category Code: Er/SIm*).* Erosion by solid particles is a problem in the aerospace, mining, agricultural, and the granular materials processing industries. Solid impingement erosion, like other forms of wear, can have functional consequences other than simply a gradual loss of material. For example, erosion of optical elements like lenses and filters can affect loss of clarity. Erosion can damage the windshields in aircraft, the smoothness of airfoil surfaces, the transmittance of sunlight into solar cells, and the attractiveness of paint finishes on automobiles. Anywhere that particles are suspended and transported by gases is likely to be affected by solid particle erosion. The medium can be flowing or the object itself can be moving through the medium. Places of particular susceptibility are located where the direction of flow of the stream changes, such as pipe elbows. Ingested particles in power turbines can erode turbine blades and guide vanes.

Erosion by slurries (Category Code: Er/Slu*).* Erosion by slurries is a major problem in material transport within the mining and chemical processing industries. ASTM G40 [6] defines a slurry as follows:

slurry, n.—a suspension of solid material in liquid.

Slurries are considered to be pumpable mixtures of solids in a liquid. Commonly, there is a chemical aspect (tribocorrosion) as well as a mechanical erosive effect associated with slurries and the pH of slurries can affect the degree to which slurry erosion takes place. Several types of slurry test methods have been developed to simulate the kinds of conditions that produce slurry erosion. ASTM test method G105-02 [46] is a wet version of G65 [11]. A second type of slurry erosion test was developed to investigate the relative abrasiveness of different kinds of slurries in the mining and petroleum industries. ASTM G75-07 [47] uses a series of lapping plates of standard reference materials to develop indices for various slurry compositions. In general, deionized water is used as the carrier fluid for the particles in the slurry. The appearance of slurry-worn surfaces depends on the direction of flow and the properties of the slurry, hence the development of standards like G75-07, which attempt to quantify the severity produced by different slurry compositions on the same material. Generally, slurry erosion is distinguished more based on where it occurs and what the wear pattern looks like (see Section 3.3) than what the surface features look like in detail.

Erosion by collapsing bubbles (also known as cavitation erosion) (Category Code: Cav). Cavitation is a process that occurs in liquids when bubbles form and collapse, producing jets (pressure pinpricks) against a solid surface. As a result, numerous pits are produced. These can grow deeper and join to produce serious damage. Sometimes, cavitation damage takes on a "frosted" appearance. Another case of frosting indicates the presence of micropitting, which is a different form of wear (see section on Contact Fatigue). Large-scale cavitation can occur on ship propellers and large fluid pumps, while other forms can occur locally around the inlets and outlets of pressurized fluid handling equipment. Figure 3.27 shows examples of large-scale and small-scale cavitation damage.

Cavitation damage from bubble collapse often has a granular or macropitted appearance on metal surfaces. Few materials are immune to the effects of cavitation damage, so component design and changing the fluid flow conditions are approaches to its minimization [48,49]. In addition to moving objects through a fluid to create a trailing bubble field, the wearing object can be stationery as a fluid moves past it. Therefore, cavitation can occur in the presence of liquid jets, and test methods have been developed to study that situation as well [50].

Spark erosion (Category Code: Er/Spk). Spark erosion occurs in obvious places such as during the throwing of electrical switches, in rotating electrical motor brushes, in sliding pantographs on electric trains, and in power relays that carry large currents. It can also occur in some less likely places such as the internal surfaces of rolling element bearings, where it can appear as craters, frosted areas, or tracks [51,52]. What is needed is a large electrical potential difference between surfaces separated by a gap. Such potentials can develop during bearing operation and need not be confined to electrical equipment. Depending on where it occurs, the evidence for spark erosion can be mistaken for micropitting and may be highly localized.

(a)

(b)

Figure 3.27 **Cavitation damage at different scales. (a) The blades of massive water turbines used at the Boulder Dam hydroelectric plant on the Columbia River in Washington State suffer from cavitation. (b) Cavitation damage on the author's coffeemaker's hot plate was caused by boiling water that got trapped under the coffee carafe. The largest flakes in image (b) are several millimeters across.**

Submerged or open liquid jet erosion (Category Code: Er/Jet-S, -O). A submerged jet (Er/Jet-S) means that the process takes place inside a fluid environment. An example is a submerged nozzle aimed at a pump impeller or vane (see also ASTM G134-95 [50]). In an open jet (Er/Jet-O), a stream of liquid impinges on a free surface. An example of open jet erosion is the cleaning of concrete or metal surfaces using a pressurized stream of water.

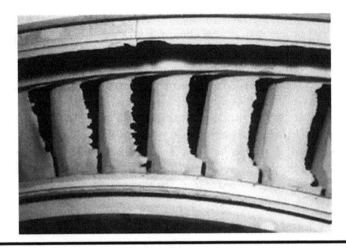

Figure 3.28   Steam erosion of the suction side of a steam power turbine showing a feathered appearance at the trailing edges of the blades. (From Martínez, A., Martínez, F., Velázquez, M., Silva, F., Mariscal, I., and Francis, J., *Energy Power Eng.***, 4, 365–371, 2012 [53]. With permission.)**

Steam erosion (Category Code: Er/Stm*).* Like streams of solid particles and liquid jets, steam flows can also lead to severe erosive wear. Surface degradation can involve not only momentum transfer from the steam but also corrosion. Therefore, steam erosion can be considered as a form of tribocorrosion or erosion–corrosion (see "Wear with Chemical Attacks" later in this chapter). This form of surface damage is common in steam turbines and guide vanes and, in extreme form, can remove most of the material in the component leaving a pitted and feathery appearance (see, for example, Figure 3.28) [53].

3.2.2.3 Contact Fatigue (Category Code: CF)

Rolling contact fatigue (Category Code: RCF/P). Those who study the dynamics of contact of rolling elements will often point out that pure rolling contact is rare and difficult to achieve. Even in a complement of balls rolling around in a bearing race, the elastic contact produces an elliptical contact patch across the mating race. Only two locations across that patch experience "pure" rolling. The area between the two rolling locations slips in an opposite direction to that outside those two points [54]. In another example, in spur gear engagements, a tooth engages the facing tooth with a measure of slip that varies with location on the teeth. It decreases until the contact point reaches the pitch circle, where an instant of pure rolling occurs before the slip increases again to a maximum value and the given tooth disengages. The amount of slip and rolling depends upon the gear geometry. Therefore, in rolling element bearings, the surface can take on a mixed-mode appearance. That is, some of the features may be due to rolling contact fatigue and others may distort those features as a measure of slip-produced tangential deformation or smearing.

Designers of rolling element bearings (radial and cylindrical), and those who use them, have commonly relied on an expression for what is called the L_{10} lifetime, and that quantity has found its way into international design standards such as ISO 281:2007 [55]. Zaretsky has critiqued this approach [56]. He recalls early work by Arvid Palmgren (1924) and later by Palmgren and Lundberg (1947); this limit is based on the probability that 90% of a given type of bearing will have survived after a calculated number of revolutions. In its simplest form, letting C_D be the dynamic load capacity of the bearing, P_{eq} be the radial load on the bearing, and p be the load-life exponent, then

$$L_{10} = \left(\frac{C_D}{P_{eq}}\right)^p.$$ (3.3)

After extensive testing under controlled conditions, a great deal of data were accumulated, and it was common to find parameters for this equation tabulated in rolling element bearing catalogs and handbooks. Later, additional correction factors were added as coefficients in Equation 3.3 to compensate for improvements in alloy processing, better cleanliness steels, contamination, compressive residual stresses, and other factors. However, bearings often fail by factors other than rolling contact fatigue (e.g., corrosion, overloads in other than the radial direction), and Zaretsky, whose recent work is described in Van Rennselar [56], pointed out that a large quantity of data for bearings made from AISI 52100 steel failed to show clear evidence of a fatigue limit. As will be noted, bearings can fail for a range of different reasons, but contrary to those observations, the use of equations like Equation 3.3 inherently presumes a priori that there is only one primary cause: rolling contact fatigue. A tribosystem analysis does not presuppose a reason for failure ahead of time but rather defines the problem and compiles the evidence and observations in preparation for a broader and more enlightened analysis.

Figure 3.29 shows an example of a battered spherical rolling element from the gearbox of a commercial wind turbine unit that was taken out of service (Ponnequin Wind Farm, northern Colorado, United States). It shows a variety of damage associated with spalling, plastic deformation, and even indentation by impact, demonstrating that a wear surface can contain artifacts from a cumulative history of what happened to the surface. What we observe is the surface condition at one point in the component's history, and the challenge is hypothesizing the sequence of events that led to that appearance.

Figure 3.29 **Severely damaged rolling element bearing from a sun gear in a wind turbine gearbox. The edges of the spalls are rounded and distorted by impact and slip as the bearing deteriorated. To the left of the center, multicolored oxide films cover the center of two of the dimples, suggesting momentary scuffing with its attendant high temperature.**

There has been a great deal of attention paid to characterizing and measuring RCF damage on rolling element bearings, and as a result numerous reviews exist (e.g., ASM Handbook [2] and Halmi and Andersson [57]). Rolling contact damage can occur at the asperity level (e.g., few tens of micrometers), in which it is called *micropitting*, or it can produce more extensive macroscale spalling in which chunks of material, easily visible to the naked eye, are lost. Micropitting (Category Code: RCF/MP) is attributed to asperity-level fatigue that produces tiny pits that can proliferate, link up, and lead to more severe contact fatigue damage. The ASTM G40-13b [6] definition is as follows:

> *micropitting*, n.—*in tribology*, a form of surface damage in rolling contacts consisting of numerous pits and associated cracks on a scale smaller than that of the Hertz elastic contact semi-width.

Micropitting at the asperity level can be the beginning of a process that leads to more severe rolling contact damage. For example, micropitting can occur on a spherical rolling bearing and eventually alter its profile (crown). That change in geometry shifts the load to one edge where macropitting or spalling or sliding wear can later develop (R. Errichello, GearTech, Montana, private conversation, 2011) [58]. Figure 3.30a,b shows close-ups of the component in Figure 3.29. They indicate that more than one form of surface damage can occur on the same part. A sketch of the corresponding outer race of the double-row spherical bearing also

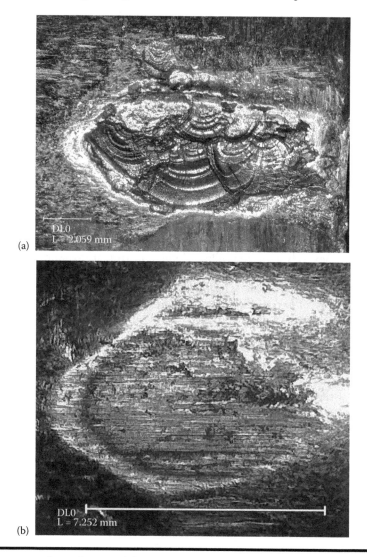

(a)

(b)

Figure 3.30 Enlarged areas of the same bearing shown in Figure 3.29. (a) Evidence of subsurface fatigue crack propagation leading to macrospalls; and (b) a large, 7.3-mm-long depression surrounded by a halo of discolored heat tints (red orange and blue in the original image) that suggests momentary elevated temperature from frictional contact.

shows the locations of the several forms of wear that occur due to differences in the motions and forces at different locations on the component (Figure 3.31). An axial overload was most likely the cause for the difference in wear between oppositely tilted races. The cross-section in the figure shows a wear step from forcing the edges of the rollers on the left side to dig in under an axial load from right to left. The right side outer race is frosted (micropitted), but the left side is more heavily spalled. Multiple forms of wear damage are possible, even on the same component.

The size of typical micropits in rolling elements is of the order of a few micrometers and less than the Hertz contact diameter, as shown in Figure 3.32 on a rolling contact test specimen of AISI 440XH stainless steel subjected to a combination of slip and rolling for a total of 4,500,000 cycles in a commercial lubricant.

Rolling with slip (Category Code: RCF/Slp). When rolling contact occurs with slip, the appearance of the wear surface can be distorted from a pitted appearance to a more smeared appearance in which some of the pits are covered over or elongated in the slip direction.

Figure 3.33 shows indications of pitting and spalling on the teeth of a hardened steel gear from an automotive transmission. Characteristically, this form of damage can be seen surrounding the pitch circle on the gear teeth. Note that for a spur gear, the amount of slip increases with distance from the pitch circle. Therefore, the slide/roll ratio for such gears also increases to a maximum at the tooth engagement and disengagement points.

Figure 3.34a and b illustrates hairline and chevron-type tensile cracking, respectively. Thin hairline microcracks have formed transverse to the slip direction, likely

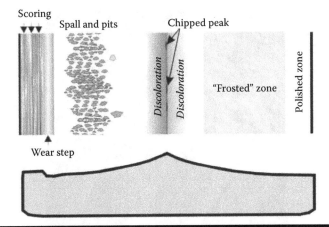

Figure 3.31 Schematic cross-section and plan view of rolling contact damage on the outer race of a wind turbine gearbox double-row spherical bearing. (From Blau, P. J., Walker, L. R., Xu, H., Parten, R., Qu, J., and Geer, T., *Wear Analysis of Wind Turbine Gearbox Bearings*, Oak Ridge National Laboratory, Technical Memo, ORNL/TM-2010/59, Oak Ridge, TN, 51 pp., 2010 [58].)

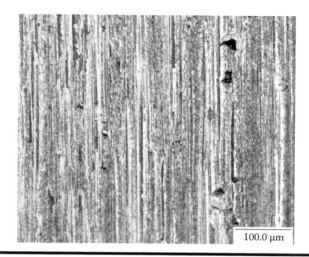

Figure 3.32 Micropitting damage on a stainless steel test specimen obtained after 4.5 million cycles of contact in a lubricated rolling/sliding test. (See also from Blau, P.J., Yamamoto, T., Cooley, K.M., and Reeves, K.S., *Investigation and Characterization of Micro-Pitting in Bearing Steels*, presented at STLE Annual Meeting, Detroit, MI, May 2013 [59].)

Figure 3.33 An example of wear type RCF/Slp on a worn steel gear. The width of the damaged areas is approximately 2–4 mm in this case.

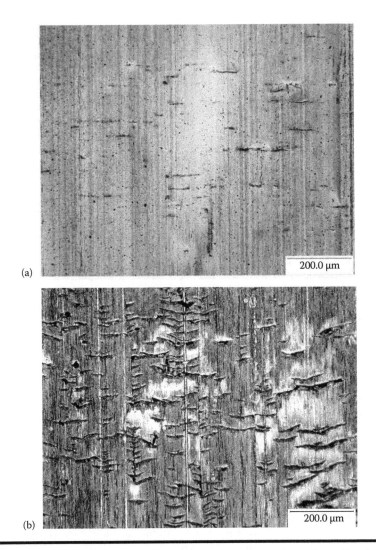

(a)

(b)

Figure 3.34 **(a) Hairline cracks forming transversely to the rolling/slip direction on a test specimen of heat-treated M50 tool steel. (b) More severe, chevron-type tensile cracks generated on a different material under the same rolling contact test conditions.**

resulting from the tensile stress fields produced behind the passing rolling element. Due to the imposed resolved stresses, such microcracks are commonly inclined to the surface by about 30 degrees (see Figure 3.35). However, complex configurations of branched cracks have also been observed in what are sometimes referred to as "white-etching layers."

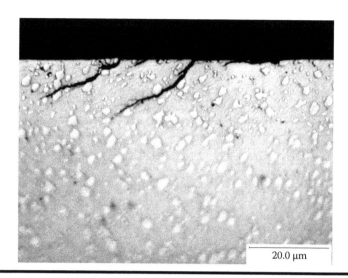

20.0 µm

Figure 3.35   A cross-section of a rolling contact fatigue specimen of Type 440HX stainless steel showing inclined tensile cracks. This is a common form of subsurface crack configuration produced during RCF.

Micropitting can occur not only with nearly pure rolling but also in the presence of slip that distorts the shapes of the micropits and exacerbates the severity of damage. Figure 3.36 shows a rolling contact test specimen of 52100 bearing steel in which the micropits that formed are distorted under the influence of slip. A number of the pits are linked and could have produced larger spalls had the test been continued.

Since the processes of contact fatigue and slip can occur simultaneously, the severity of the damage and the array of features that exhibit those forms of wear can be complicated. Figure 3.37 summarizes the concept that a combination of wear processes can occur simultaneously on rolling elements and at various levels of contact severity (high contact pressures, fatigue, and shear tractions). Four levels of pitting damage and four levels of sliding damage have been proposed [59]. P refers to pitting-related processes, and S, to slip- or sliding-related processes. For example, Figure 3.36a would be considered P1, Figure 3.36b would be considered an example of level P3, and Figure 3.30a is an example of P4.

Normal vibrational contact (sometimes called "Hertzian contact fatigue") (Category Code: -VC). If two bodies remain in contact as the normal load varies, the material below the contact may begin to accumulate damage, initiate cracks, grow them, link them, and generate loose debris. Figure 3.21 showed a simplified schematic of a VC arrangement that can also be considered radial fretting depending on the amplitude and degree of slip. As the nonconformal upper body oscillates under load range ΔP, the contact diameter (2a) expands and contracts accordingly. The

Figure 3.36 Severe micropitting damage on AISI 52100 steel rolling with 5% slip. Here, the larger pits are linked and partly smeared over. (From Blau, P.J., Yamamoto, T., Cooley, K.M., and Reeves, K.S., *Investigation and Characterization of Micro-Pitting in Bearing Steels*, presented at STLE Annual Meeting, Detroit, MI, May 2013 [59].)

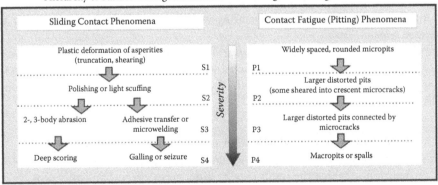

Figure 3.37 Summary of parallel paths that can produce different combinations of damage features depending on the combination of slip and rolling contact fatigue. (From Blau, P.J., Yamamoto, T., Cooley, K.M., and Reeves, K.S., *Investigation and Characterization of Micro-Pitting in Bearing Steels*, presented at STLE Annual Meeting, Detroit, MI, May 2013 [59].)

material below the contact experiences contact fatigue, while the outer annulus of the contact at the surface can experience short amplitude radial sliding (a special case of radial fretting) as the expansions and contractions take place.

The strict connotation of Hertzian contact involves elastic response, so if the material within that area is deformed plastically in the process, one cannot strictly refer to this wear form as Hertzian. Yet, calling it Hertzian contact fatigue conveys the message that there is vertical, oscillatory motion leading to surface distress. The appearance of such damage, especially in hard, brittle materials like some ceramics, may include circumferential cracks ("ring cracks"), radial surface-breaking cracks, or "cone cracks" that expand out below the surface. Eventually, clam shell chips may appear at the edges of the contact region.

3.2.2.4 Wear with Chemical Attacks

Mechanical contact often occurs in the presence of chemical environments. The two can have synergistic effects to either enhance or reduce surface damage relative to the occurrence of chemical attack or mechanical attack by themselves. For example, shot peening (mechanical working by repetitive impact of shot particles) can improve corrosion resistance. Scratched surfaces of aluminum can resist exfoliation corrosion.

Chemomechanical polishing wear (Category Code: CMPol*).* As mentioned in the discussion on abrasive wear, a polishing process can range from mainly chemical polishing, with little or no mechanical contribution, to a more grinding form of polishing that involves 2-body or 3-body abrasion. When there is clearly a combination of both actions, the term *chemomechanical polishing wear* can be used. More broadly, and not necessarily connected with polishing, when mechanical wear is either reduced or enhanced by operating within a chemical environment, this is termed *tribocorrosion*.

Tribocorrosion (Category Code/Suffix: TrCor*).* When both wear and chemical corrosive attack occur together in such a way as to alter the wear rate, then tribocorrosion is indicated. ASTM standard G40-13b [6] defines it as follows (without the hyphen):

> *tribocorrosion*, n.—a form of solid surface alteration that involves the joint action of relatively moving mechanical contact with chemical reaction in which the result may be different in effect than either process acting separately. Synonym: *wear-corrosion synergism*.

The implication from the definitions of the term *tribocorrosion* is that wear and corrosion occur simultaneously, but that is not always the case. They may also occur in sequence to produce the net effect of material load and surface degradation. Table 3.3 lists a few examples.

Table 3.3 Sequential or Simultaneous Tribocorrosion

Sequence	Effect	Examples
Corrosion followed by wear	Corrosion products (scales) can be loosened and then removed by subsequent mechanical contact.	Rusted brake drums or rotors from standing in the rain are then subjected to braking; bearings that rust on the shelf before being placed in service.
Corrosion with wear	Corrosion in conjunction with mechanical contact produces rates of loss different than either separately.	Wet industrial processes; wear of lubricated engine parts; human body implants; slurry erosion in mining.
Wear followed by corrosion	Rubbing removes a passive or protective film or coating that enables corrosion of the exposed substrate.	Corrosion of worn-out mining equipment; corrosion of highway guard rails that have been scuffed by accidental contact with a vehicle.

Tribocorrosion takes place in the context of any number of types of relative motion like sliding, erosion, or abrasion, so simply identifying the process as TrCor does not sufficiently identify the contact condition. Therefore, the code is used as a suffix with other category codes to indicate the strong contribution of corrosion to the surface damage process. For example, SNR/TrCor identifies the case where reciprocating sliding occurs in a chemically active environment.

The study of tribocorrosion has seen a growing interest in recent years with the publication of several books [60,61] and the introduction of a new journal (*Journal of Bio- and Tribo-Corrosion*, Springer Publications). Dental research has used the term *tribochemical* wear as a synonym for "dental erosion," emphasizing the interplay of erosion, 3-body abrasion, and attrition in tooth wear. A review of methods to apply such combinations was published by Lambrechts et al. [62]. These authors point out that the apparati used to simulate the tooth/tooth, tooth/restorative, and restorative/restorative tribosystems have taken a variety of forms, some of which simulate multiple forms of wear, and some, more simple geometries and motions.

Experimental approaches to investigating tribocorrosion sometimes involve a combination of electrochemistry experiments and wear testing (e.g., standard ASTM G119), so the complexity of the methodology and the interpretation of results have led to spirited discussions as to the relevance of test data to field situations whose environment tends to be less well controlled. In some cases, mechanical impact can retard corrosive attack (such as the use of shot peening of metals to impart corrosion resistance), and in other cases, abrasion can activate a surface and facilitate environmental attack (such as when wear removes the normally protective chromium oxide film on a stainless steel and causes it to corrode more than would otherwise be expected). Therefore, it cannot be said that tribocorrosion *always* accelerates wear processes, but only that it has a potential to alter the rate of wear relative to what is observed by mechanical processes alone.

3.3 Material Dependence of Wear Patterns and Surface Appearance

Chapter 3 has been concerned with identifying a wide range of wear types based on the type of relative motion and the kinds of features that can be observed. From a metallurgical and materials science point of view, a given physical contact situation can produce different responses from different types of materials such as pure metals, alloys, ceramics, composites, elastomers, and so on. Therefore, one cannot completely specify the general characteristics of different types of wear surfaces without some consideration of the difference in properties of the interacting materials. In abrasive situations, as was noted earlier in this chapter, an abradant could remain intact (low-stress abrasion) or it could fracture (high-stress abrasion). One would expect to see different kinds of features depending on which of these things happen. Much of the basic work in wear has been done in metals, so review articles, unless specifically focused on certain classes of materials, tend to draw strongly on the history of research based on metallic wear.

In addition to identifying the type of wear, it is also important to recognize wear patterns. This principle is widely used in the diagnosis of suspension and alignment problems from the appearance of treads on motor vehicle tires. Patterns of wear can indicate a variety of conditions of usage and potential problems with the vehicle suspension. Some tires are designed for tens of thousands of kilometers of repeated use, but others are used only once. Figure 3.38a shows the tread and shoulder of a radial design automobile tire after 45,584 miles (73,360 km) of use. Plateaus are called "tread blocks," which they contain sipes, which are small slits that reduce the shear during running and help lower the tire temperature. By contrast, Figure 3.38b shows the worn area on a tire tread from the U.S. Space Shuttle *Discovery* mission STS-121, which landed only once and traveled about 2 km.

Table 3.4 lists some of the conditions and causes for automobile and truck tire tread wear patterns. Such wear patterns can indicate wear or alignment problems in suspension system components.

Sometimes, a wear pattern can change shape as the type of wear changes. An example can be found in the previous example of spherical rolling element bearings

Figure 3.38 **(a) Wear on a radial design an automobile tire after traveling 73,360 km. (b) Wear indications on a tire used to land the U.S. Space Shuttle** *Discovery* **after traveling about 2 km. The coin is a U.S. quarter, 24.3 mm in diameter. (From Blau, P.J.,** *Friction and Wear Transitions of Materials: Break-in, Run-in, Wear-in,* **Noyes Publishing, Park Ridge, NJ, 1989 [63].)**

in which micropitting slowly changes the curvature near the equator of the barrel-shaped rollers. The contact location gradually shifts from the equatorial region of the roller to its outer edge. Increased stress concentration at that outer edge then leads to macrospalling. Therefore, one form of wear can change the shape of a component, which leads to another form of wear.

Table 3.4   Conditions Indicated by Tire Wear Patterns

Appearance	*Possible Cause(s)[a]*
Loss of tread in the center (bald around the circumference)	a. Consistent overinflation b. Wide tires on narrow rims
Loss of tread on both sides of the center circumferential portion of the tire which seems to have less wear ("shoulder wear")	a. Consistent underinflation b. Bent steering or idler arms
Tread ribs develop rounding on one side and a sharp edge on the other ("feathering"), usually on the front end	a. Toe-in set incorrectly b. Bad front suspension bushings
Worn on one side of the tire more than the other	a. Wheel alignment b. Wear in the ball joints, control arms, or sagging springs
Periodic scallops near the outer edge of the tread near the side casing ("cupping")	a. Out of balance due to worn suspension parts (ball joints, shock absorbers, springs, bushings) that cannot be cured by realignment alone b. Lack of rotation to another wheel end
Wear of the second rib in radial tires creates two bald areas about equidistant from the centerline, and some of the outer circumference less worn	a. Tire may be oversized relative to the wheel size
Outer edge of the tire wears more than the inner edge ("camber wear")	a. Alignment problem
Flat, bald spots ("patch wear")	a. Wheel out of balance b. Skidding the tires by hard braking
Cracks in treads	a. Underinflation b. Excessive speed

[a] http://www.procarcare.com/includes/content/resourcecenter/encyclopedia/ch25/25readtirewear.html; http://www.autozone.com.

As a final example, wear patterning can provide insights into other than machine diagnostics. Anthropologists have long used the patterns observed on tools and teeth to infer the daily behavior of primitive societies. A recent report [64] describes how the study of Neanderthal incisors (teeth) in Western Europe supported the different roles for males and females in a society that existed as long as 50,000 years ago. Different patterns on adult male teeth and female teeth indicated that males had more chipping and fracturing on their upper teeth and women had more on their lower teeth. Striations on women's teeth were longer and more numerous that those on men's teeth. Researchers propose that the differences indicate that men and women engaged in different tasks. For example, women may have been responsible for preparing furs and garments, and the two sexes engaged in different foraging tasks. While Neanderthals are not a subspecies of *Homo sapiens*, the longstanding assertion that the sexual division of labor was a unique characteristic of *Homo sapiens* societies is in doubt.

A summary of the types of wear and their codes was provided earlier in this chapter, in Figure 3.3. It can be used as a shorthand resource for indicating the dominant form(s) of wear in the tribosystem analysis form described in Chapter 5.

3.4 Wear Transitions

Chapter 3 has focused on the identification and categorization of the various forms of wear and related surface damage. However, the appearance of a surface at a given time is the result of its history of use. It reflects the cumulative effects of interacting with other surfaces or substances within the surrounding environment. Without additional understanding of the operating conditions and by applying a tribosystem analysis, it is difficult to know if the surface reached its present appearance in a smooth and continuous manner or in stages. For example, a surface may gradually wear-in, then the balance of competing processes will change, leading to a different form of damage and a different rate. Therefore, wear diagnosis should reflect a consideration not only of the evidence for the current wear type but also of the stages through which the surface may have passed in order to reach the point at which it is examined. Similar to the process of fatigue crack initiation, growth, coalescence, and eventual component fracture, wear life can behave stochastically. Since the cumulative exposure time to reach a given state of wear can vary even in the same tribosystem, it is advisable when possible to examine worn parts exposed for similar times as well as for various lengths of time.

Associating a specific type of wear and wear rate to a given tribosystem may be convenient for damage diagnosis and wear forecasting; the truth is that many, if not most, real engineering tribosystems do not operate under steady-state conditions [65]. As a result, wear damage can build up incrementally rather than smoothly and continuously. Unlike the aforementioned piston that sees periodic changes in

Table 3.5 Wear Scenarios and Conditions

Terms	Context
Break-in, running-in, wear-in	Changes in the friction characteristics and/or wear rate associated with the transition from as-finished surfaces to those that have been subjected to progressive contact in a tribosystem
Normal wear	Progressive wear, usually with a linear wear rate, that is expected based on the given design or predicted part replacement interval
Abusive wear	Wear caused by operating a tribosystem under harsh conditions that exceed those expected for normal operation and design parameters
Abnormal wear—"infant mortality"	Unexpected early failure relative to the design lifetime; many possible causes including improper part installation, substitution of improper materials, defects in a bearing material; bearing surface alignment errors; improper cleaning or residual manufacturing debris in a new part; unexpected start-up transients
Abnormal wear—reduced life	Intermittent high wear due to temporary deviations from the normal operating conditions, tending to reduce overall wear life beyond that for normal wear conditions; examples include wear from cold starts or unexpected starts and stops rather than continuous operation of a tribosystem
Abnormal wear—catastrophic failure	Unexpected failure during the post-run-in stage due to a single event (grit contamination, overheating event, abnormal vibration, etc.)
Wear-out	Higher-than-normal wear or surface damage that occurs at the end of the useful life of a tribocomponent
Cyclic wear behavior	While difficult to observe except in the laboratory, some surfaces may roughen, then smooth (heal), then roughen again periodically

contact conditions with each stroke, there are other systems in which operating conditions are changed intentionally by the operator (e.g., the driver braking a car) or by design (turbochargers or pumps that start up or shut down automatically depending on the power demand). Therefore, wear may occur in a cumulative manner after a spectrum of loads, speeds, temperatures, and other mechanical factors (e.g., see the study reported in Miranda and Ramalho [66]). This subject has been further discussed (e.g., Blau [63]), but more research is needed in the complex area of non-steady-state wear.

Practicing engineers are well aware of transitions in wear behavior and have developed terms to describe it. An obvious one is the change in wear rate when a protective coating wears through. Table 3.5 lists some terms that have been used to describe changes in wear or regimes of wear in sliding contact.

Some transitions in wear and friction are caused by changes in the regime of lubrication, lubricant aging (e.g., acidification or oxidation), lubricant starvation or loss, viscosity breakdown, additive package depletion, or contamination of lubricants by hard particles due to an ineffective lubricant filtering process. Additional discussion of this aspect of tribosystem operation may be found in Chapter 2 on wear detection and in Chapter 5 in the discussion of Tribosystem Analysis form, Block 2.7, that concerns the regime of lubrication.

Accounting for wear transitions, sequential steady-state stages, and examples of transitions in different forms of wear was exemplified by the author by developing a model for the fretting of nuclear power plant core components [5]. An earlier book by the author was devoted to running-in and other transitions in friction and wear [63].

3.5 Wear Diagnosis by Multiple Attributes in Context

In the author's opinion, the best way to identify the dominant form of wear on the surface of a damaged part is to use what we shall call the method of Multiple Attributes in Context (MAC). MAC is based on a careful examination of the surface damage at various magnifications with several complimentary imaging tools, studying the debris (if any can be collected), looking at both sides of the contacting couple (if sliding, rolling, or 2-body contact is involved), considering the lubricant condition, and placing these observations into context of the relative motion, contact pressure, rate of relative motion, and surrounding environment (including not only temperature and chemistry but also the dynamics of the structure). We have learned from the previous discussion that similar features can occur even in different forms of wear. This is because the same mechanisms and processes can be operating on a surface but at different degrees of severity for different wear situations. All available observations and evidence should be considered.

Using a MAC approach allows one to combine observations with the knowledge of how a tribosystem operates, as will be documented in Chapter 5, to arrive at a determination of the major form of wear on a case-by-case basis.

Consider, for example, the surface of the worn metal shown in Figure 3.16. One might immediately note the prominent, continuous grooves and draw the premature conclusion that abrasive wear was the dominant form. However, on further study, we also observe a number of torn, smeared, ductile features. Additional

Table 3.6 Features Common to Several Types of Wear (Metallic Surfaces)[a]

Feature	2BAb	3BAb	S/NU	S/NR-Fr	Er-Slm	RCF-Slp
Ductile shear, tearing	C		P	M		
Transfer of material to the counterface			C	C		
Fine striae parallel to the motion direction	P		C			C
Deep grooves parallel to the motion direction	P	M	C			
Skewed indentations, curved digs, or grooves	M	C	M		M	M
Flat, shiny debris flakes			P			
Cutting chip-like debris	C	C	M			
Fine powdery debris, typically oxidized			M	P		
Dull, matted appearance				C	P	C
Fine-scale pits				C	P	C
Macroscale spalls						P
Multidirectional scratches		P				
Shiny or specular appearance (polishing)	C	C	M			M
Halo of damage in concentric zones				P		
Embedded particles				M	C	
Chevron-type microcracks transverse to the direction of motion	M		M			M
Delamination of surface layers			C	C		M

Note: C, common, but not always observed; M, possible, but less common; P, prominent, common feature.

[a] This table compares the appearance and occurrence of some types of wear surface and debris features. It does not purport to indicate subtle or detailed differences and is not intended to be all-inclusive.

study reveals the presence of a few adhered flakes of metal such as that at the upper right of the image and about two-thirds the distance across the bottom edge. The large metallic wear flakes shown in Figure 3.16 (as opposed to curled chip-like flakes that occur with abrasive, cutting wear) and a patch of material adhered to the counterface in that experiment suggested the presence of adhesive wear. Work-hardening of asperities and debris fragments during sliding can make the affected metal hard enough to plow the opposing surface and by that process produce features akin to those for 2- or 3-body abrasive wear. Therefore, there is a commonality of mechanisms and processes in abrasive and adhesive wear. Yet, when we observe some other attributes of the situation—such as metallic flakes and adhered material on the counterface—it becomes clearer that adhesive wear dominated.

In another example, the same MAC approach can be used to identify fretting wear with tribocorrosion. The surface of a dry steel contact surface, say a bolted joint subjected to vibration, when opened, may reveal a localized region of granular-looking pits surrounded by finely powdered, reddish debris (historically called red mud). Granular surface damage plus fine oxidized debris points to fretting. Given additional information on low-amplitude oscillations to which the joint was subjected, fretting wear is further confirmed.

Table 3.6 illustrates how several different forms of wear can have similar-appearing surface features. Seeing one type of feature may be a clue but may not uniquely identify a specific form of wear. Surface and debris observations, coupled with a knowledge of the operating conditions, like those documented within the tribosystem analysis in Chapter 5, are basic to the MAC approach.

Appendix 3A: Wear Nomenclature and Nomenclature Families

A wide variety of terms have been used to describe wear and associated surface damage. Bayer [A1] (pp. 159–161), for example, lists 87 mechanistic classifications of wear mechanisms, and elsewhere in his book (p. 30), another 38 more terms were used to describe "wear behavior and mechanisms." There is no doubt that the proliferation of nomenclature is a confusing factor in the tribology literature, and it can also confuse wear diagnosis and searches for similar cases based on "key words."

In this appendix, a number of wear-related terms have been collected and their context is noted. Formal definitions for a number of these terms may be found in ASTM G40-13(b) [A2], Blau [A3], and Winer and Peterson [A4]. Other terms (the jargon of tribology) have been coined in the field to describe what is observed on load-bearing surfaces. As a result, there is a lot of ambiguity in the application of wear terminology, but some terms are in widespread use despite their imprecision.

Note that some terms refer to higher forms of wear (e.g., abrasive wear), others to wear processes (e.g., plowing), and still others to fundamental mechanistic interactions (e.g., adhesion).

Part of the problem of terminology usage in the tribology literature is the use of the same terms when describing different types of wear phenomena. For example, the term *plowing* (spelled as *ploughing* in the United Kingdom) can be used to describe the result of a hard abrasive point (2-body abrasive wear) moving through a softer counter-surface. Likewise, it can also be used to describe the digging in of loose abrasives in the case of 3-body abrasion or particle erosion. Plowing has also been used to describe one of the two components of metallic friction: a plowing component and an adhesive component.

To help sort out the use of terms in tribology, one can define *nomenclature families*. A nomenclature family is a group of terms commonly associated with a certain form of wear. A given term may be unique to that form of wear or it may be shared with other forms of wear. For example, 2-body wear uniquely describes a certain form of wear, but the term *plowing*, as discussed previously, is associated with other forms of wear and frictional phenomena. Figure 3A.1 schematically shows the abrasive wear nomenclature family.

The list of nomenclature in Table 3A.1 has been annotated to refer to sections in this book where they are described or to one or more references from which they are defined. The purpose of column 2 of the listing ("Context") is not to provide formal definitions of these terms, but rather to explain where they tend to be used and how.

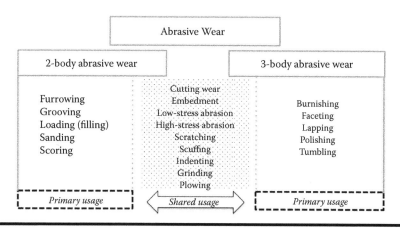

Figure 3A.1   Some members of the abrasive wear family of nomenclature. Terms that are more directly associated with 2- and 3-body abrasion are shown at the left and right, respectively, while some terms shared between these types and sometimes with other forms of surface damage are shown in the middle.

Table 3A.1 Nomenclature Used in the Wear Literature

Term	Context	Applicable Wear Codes	Appendix Ref. No.
Abradant	Referring to that which causes abrasion		[A2]
Abrasion	A process of removing material by hard asperities or loose particles		
Abrasive	May be used as a synonym for abradant; used in conjunction with grinding or polishing media; also descriptive of a process (adj.)		
Abrasive wear	The result of abrasion (see also 2- and 3-body abrasive wear)	S/Ab, S/2BAb, S/3BAb, S/2-3Ab	[A2]
Adhesive wear	Historical description of a severe wear process involving material transfer, ductile grooving, flake-like debris	S/–	[A1], [A3]
Asperities	High spots on a contact surface; term is generally associated with wear and friction research and modeling		
Attritious wear	Wear of abrasive grains on grinding wheels	S/Ab	
Beach marks	Indications of fatigue wear processes observed on spalled surfaces		[A5]
Brinelling	Vibratory formation of shallow elliptical indentations that retain the original surface finish		[A2]
Burning	Color-tinted areas suggestive of overheating		[A3]
Burnishing	Usually an industrial process to apply a substance to a surface by rubbing		

(Continued)

Table 3A.1 (Continued) Nomenclature Used in the Wear Literature

Term	Context	Applicable Wear Codes	Appendix Ref. No.
Cavitation	A process by which collapsing bubbles in a liquid impinge on a surface, resulting in tiny pressure spikes		[A2]
Cavitation erosion	Wear caused by the process of cavitation		[A2]
Chafing	General term may involve abrasion or nonabrasive rubbing		[A3]
Chatter marks	Periodic markings on a machined or sliding surfaces generally associated with vibrations or high-frequency instabilities in contact conditions		
Checking	See "craze cracking"		[A3]
Chemical wear	Material removal by wear but assisted by active chemical species	TrCor	
Chevrons	Tent-like wear features that are sometimes attributed to tensile cracking of brittle materials		[A5]
Chipping	Creation of nicks or small areas of material removal; tends to be associated with brittle material		
Cocoa	A type of fretting wear debris		[A3]
Contact fatigue	A process that produces surface damage by the repetitive application of a normal force to a region of the surface	CF	
Corrosive wear	Mechanical wear in which corrosion also plays a part in material removal; see also tribocorrosion		
Crater wear	A form of cutting tool wear in which shallow craters are formed on the rake face of the tool		

(*Continued*)

Table 3A.1 (Continued) Nomenclature Used in the Wear Literature

Term	Context	Applicable Wear Codes	Appendix Ref. No.
Craze cracking	Also known as "map cracking" in reference to concrete; associated with thermal cycling in metals; also called heat checking on face seal surfaces		
Delamination (wear)	A wear process in which debris particles shaped like flat platelets are produced by fatigue crack growth, similar to peeling	(Associated with S/N)	
Diffusional wear	Heat-induced changes in surface chemistry and properties usually associated with wear of machine tool inserts		
Dusting	Usually associated with the powdery deterioration of electric motor brushes or current carrying contacts		[A3]
Erodant	Particles that are capable of causing erosive wear		[A2]
Erosion	A process by which multiple particles impinge on a surface		[A2]
Erosive wear	General type of wear caused by erosive processes that can involve different kinds of erodants	Er/LDr, Er/Slm, Er/Slu, Er/Cav, Er/Spk, Er/Jet-S, Er/Jet-O	
False brinelling	Indentations on a bearing race that resemble brinelling, but do not retain the original surface finish in those dimple marks		[A2]
Fatigue wear	Wear due to repetitive contact; the direction may be tangential, normal, or at an arbitrary angle to the surfaces		[A2]

(Continued)

Table 3A.1 (Continued) Nomenclature Used in the Wear Literature

Term	Context	Applicable Wear Codes	Appendix Ref. No.
Flaking	Formation of loose, flat particles or debris		[A3]
Flank wear	Wear of a machine tool that occurs on the face of the tool that slides along the cut surface (as opposed to rake face wear)		[A3]
Flow cavitation	Cavitation produced by the movement of a fluid past a solid object		[A2], [A3]
Fluting	Periodic grooves on a bearing surface, sometimes associated with electrical burns		[A3]
Fretting	Low-amplitude oscillatory motion between surfaces		[A2]
Fretting wear	Wear cased by fretting motion	S/NR-Fr	[A2]
Fretting corrosion	Early term for fretting due to its production of oxidized wear debris		[A2]
Fretting fatigue	Fretting motion that causes the nucleation and propagation of cracks		
Frosting	Used in bearing tribology, generally a matte appearance of a bearing surface due to an area of micropitting		[A3]
Furrowing	Producing deep grooves in a wear surface		
Galling	A form of surface damage associated with severe plastic deformation that may lead to lumps, seizures, or the inability of surfaces to mate properly		[A2]

(Continued)

Table 3A.1 (Continued) Nomenclature Used in the Wear Literature

Term	Context	Applicable Wear Codes	Appendix Ref. No.
Glazing	Formation of a tribofilm on one or both rubbing surfaces; may be due to the detachment of material from opposing surfaces and their compression or mixing into a distinct third-body layer		
Gouging	Used in conjunction with abrasion; the formation of deep grooves usually in metal surfaces; examples are mining and ore-crushing equipment		
Grinding abrasion	Combination of abrasive and impact wear common in rock-crushing operations		[A6]
Grooving	Formation of relatively long, deep, continuous ductile grooves; severe form of 2-body abrasion		
Heat checking	Formation of cracks on burn marks on a sliding surface, commonly associated with face seals or other flat/flat contacts at high speeds or depleted lubrication		
Hertzian contact fatigue	Nucleation and growth of subsurface cracks due to a oscillating normal force on a surface, usually associated with bearing failures; may involve rolling or vertical oscillation	RCF/VC	[A2]
High-stress abrasion	Abrasion in which the load is so severe that the abradant particles fracture		
Hollow wear	Usually associated with railroad wheels; wear depth below the flanges of the wheels		[A7]

(Continued)

Table 3A.1 (Continued) Nomenclature Used in the Wear Literature

Term	Context	Applicable Wear Codes	Appendix Ref. No.
Impact abrasion	Combination of impact motion with abrasion; may be caused by inclined impact of two surfaces or by hammering in the presence of abradants		
Impact wear	Wear caused by repetitive impact with an opposing surface or loose particles; sometimes called hammering	RI/2B, RI/3B	[A1], [A2]
Impingement erosion	Erosion by a flux of particles (liquid, solid, or mixtures)		[A2]
Jet erosion	Also known as liquid impingement erosion		[A2]
Kinematic wear marks	Short, arc-shaped scratches on bearing surfaces arising from trapped hard particles		[A3]
Lapping	Removal of material, usually by 3-body abrasion		
Low-stress abrasion	Abrasion process in which shallow, fine grooves are produced and the abradant particles remain intact		[A3]
Macropitting	In wear, the formation and loss of spalls generally visible to the naked eye; common indication of fatigue wear		[A5]
Magic wear	In railroad track wear, a condition in which the continuous wear removes near-surface fatigue cracks and thus avoids fatigue crack propagation		[A8]
Micropitting	Fine-scale pitting similar in size to grains or asperities		[A2]

(Continued)

Table 3A.1 (Continued)   Nomenclature Used in the Wear Literature

Term	Context	Applicable Wear Codes	Appendix Ref. No.
Mild wear	Sometimes called oxidational wear in metals, produces fine debris without significant plastic deformation or major loss of material		
Nonabrasive wear	General category used by ASTM to differentiate sliding wear from wear dominated by abrasion; includes adhesive wear, scuffing, galling, plastic smearing	S/N, S/NU, S/NR, S/MD/-, S/NR-GS, S/NR-Fr	
Normal wear	Vague term that reflects the expected wear rate and associated processes in specific tribosystems (as opposed to abnormal or high wear)		
Notching wear	Grooves that form on machining inserts		
Oxidative wear	A mild form of wear in which oxide scales are formed and debris is released from those oxidized layers		[A2]
Peening wear	Repetitive impact		[A3]
Plowing (United Kingdom: ploughing)	A process in which one or both wear surfaces are grooved as by a plow		[A2]
Polishing wear	A very fine-scale removal of material that may involve a mild abrasion, chemical attack, or both	S/Ab/Pol; CMPol	[A5]
Rain erosion	Erosive wear from impinging rain particles		[A3]
Rake face wear	Wear on the secondary shear zone of a cutting tool where the cut chip slides along it (related to crater wear)		

(Continued)

Table 3A.1 (Continued)   Nomenclature Used in the Wear Literature

Term	Context	Applicable Wear Codes	Appendix Ref. No.
Red mud	Fine wear debris associated with fretting; see also cocoa		[A3]
Ridging	Referring to the formation of ridge-like features, usually in bearings or gears		[A3], [A5]
Roll formation	In sliding contact, the formation of needle-like particles usually found lying perpendicular to the direction of relative motion		
Rolling contact fatigue	A form of fatigue wear due to relative rolling motion, often involves slip plus rolling	RCF, RCF/P, RCF/MP, RCF/Slp	[A1]
Scoring	Similar to grooving, generation of grooves mainly by plowing		[A2]
Scratching abrasion	Somewhat ambiguous terminology associated with 3-body abrasive wear		[A6]
Scuffing	Damage associated with the sliding of materials often attributed to lubricant failure or starvation; it may roughen or smooth the surfaces, but plastic deformation is always involved		[A2]
Severe (metallic) wear	Roughening and material loss from a metal surface associated with plowing, plastic deformation, and transfer; commonly produces shiny flake-like wear particles	S/U or S/R	
Shelling	Advanced form of spalling		[A3]
Seizure	Ceasing of relative motion due to adhesion or interlocking of surface features		

(Continued)

Table 3A.1 (Continued)   Nomenclature Used in the Wear Literature

Term	Context	Applicable Wear Codes	Appendix Ref. No.
Sliding wear	General category of tangential motion; may involve abrasive and/or adhesive wear; may be unidirectional or reciprocating or multipath motions		[A1]
Slurry erosion	Erosion by a mixture of particles suspended in a liquid		
Smearing	Plastic deformation of a contact surface so as to produce thin layers, as a result of tangential motion		[A3]
Spalling	Loss of irregular fragments of material; a product of fatigue processes		[A2], [A3]
Spark erosion	Removal of material by arcing electrical sparks		[A3]
Stick slip	A periodic seizure and release of surfaces that can produce vibrations and chatter; unstable condition of sliding		[A3]
Surface distress	Production of a thin, work-hardened surface layer by plastic deformation; commonly used in conjunction with bearings		[A3]
Tensile cracking	In sliding contact, the production of fine surface-breaking microcracks perpendicular to the sliding direction and inclined to the surface		
Three-body abrasion	Form of abrasion involving loose particles (abradants)	3BAb	
Transfer	A process by which sliding materials like clean metals tend to adhere to a rubbing partner		[A3]

(Continued)

Table 3A.1 (Continued)   Nomenclature Used in the Wear Literature

Term	Context	Applicable Wear Codes	Appendix Ref. No.
Tribocorrosion	Combined action of mechanical wear chemical attack	TrCor	[A2]
Two-body abrasion	Abrasion caused by sliding between a solid surface and a counterface of fixed, hard points	2BAb	[A2]
Vibratory cavitation	Alternate way to express cavitation erosion		[A3]
Wear	General term for periodic loss or displacement of material by mechanical action		[A2]

References

A1. R. G. Bayer (2002) *Wear Analysis for Engineers*, HNB Publishing, New York, pp. 30 and 159, Chapter 3.1.

A2. ASTM G40-13(b) (2014) "Terminology relating to wear and erosion," in *Annual Book of Standards*, Volume 03.02, ASTM International, W. Conshohocken, PA, pp. 160–168.

A3. P. J. Blau, ed. (1992) "Glossary of friction, wear, and lubrication terms," Vol. 18, *Friction, Lubrication, and Wear Technology*, ASM International, Materials Park, OH.

A4. W. O. Winer and M. B. Peterson, eds. (1980) *Wear Control Handbook*, ASME, New York.

A5. R. Errichello (2015) "Failure analysis glossary," Geartech, www.GearBoxFailure.com.

A6. H. S. Avery (1961) "The measurement of wear resistance," *Wear*, Vol. 4, pp. 427–449.

A7. K. Sawley, K. Urban, and R. Walker (2005) "The effect of hollow-worn wheels on vehicle stability in straight track," *Wear*, Vol. 258, pp. 1100–1108.

A8. E. Magel, M. Roney, J. Kalousek, and P. Sroba (2003) "The blending of theory and practice in modern rail grinding," *Fatigue and Fracture*, Vol. 26, pp. 921–929.

Appendix 3B: Resources for the Analysis of Wear Problems

3B.1 Images of Wear Surfaces and Failed Components

M. J. Neale, ed. (1995) *Component Failures Maintenance and Repair: A Tribology Handbook*, Butterworth-Heinemann, Oxford, United Kingdom.

R. G. Bayer (2002) *Wear Analysis for Engineers*, HNB Publishing, New York.

K. G. Budinski (1988) *Surface Engineering for Wear Resistance*, Prentice Hall, Englewood Cliffs, NJ.

K. G. Budinski (2013) *Friction, Wear, and Erosion Atlas*, CRC Press/Taylor & Francis Press, Boca Raton, FL.

T. E. Tallian (2000) *Failure Atlas for Hertz Contact Machine Elements*, ASME, New York.

3B.2 Wear Debris Appearance

Wear Particle Atlas (1982) Naval Air Engineering Center, Report NAEC-92-163, Naval Air Systems Command Final Technical Report.

B. J. Roylance and T. M. Hunt (1999) *The Wear Debris Analysis Handbook*, Coxmoor's Machine and Systems Condition Monitoring Series, Coxmoor, United Kingdom.

3B.3 Oil Analysis and Functions of Additives

D. Troyer and J. Fitch (1999) *Oil Analysis Basics*, Noria Corporation, Tulsa, OK.

L. R. Rudnick, ed. (2009) *Lubricant Additives—Chemistry and Applications*, 2nd ed., CRC Press, Boca Raton, FL.

References

1. H. S. Avery and the ASM Committee on Wear Failures (1975) "Wear failures," in ASM Handbook, *Failure Analysis and Prevention*, Vol. 10, 8th ed., H. E. Boyer, ed., ASM International, Materials Park, OH, pp. 134–153.
2. ASM Committee on Wear of Sliding and Rolling Element Bearings (1975) "Failures of rolling element bearings," in ASM Handbook, *Failure Analysis and Prevention*, Vol. 10, 8th ed., H. E. Boyer, ed., ASM International, Materials Park, OH, pp. 416–437.
3. ASM Committee on Gear Failures (1975) "Failures of gears," in ASM Handbook, *Failure Analysis and Prevention*, Vol. 10, 8th ed., H. E. Boyer, ed., ASM International, Materials Park, OH, pp. 507–524.
4. P. J. Blau (1994) "The terminology of friction, lubrication, and wear," in *Tribology in the USA and the Former Soviet Union*, V. A. Belyi, K. C. Ludema, and N. K. Myshkin, eds., Allerton Press, New York, pp. 431–442. Published in Russian and English versions.
5. P. J. Blau (2014) "A multi-stage wear model for grid-to-rod fretting of nuclear fuel rods," *Wear*, Vol. 313 (1–2), pp. 89–96.
6. ASTM G40-13b (2014) "Terminology relating to wear and erosion," in *ASTM Annual Book of Standards*, Vol. 03.02, ASTM International, W. Conshohocken, PA, pp. 160–168.
7. P. J. Blau (2009) "Embedding wear models into friction models," *Tribology Letters*, Vol. 34 (1), pp. 75–79.

8. P. J. Blau (2008) *Friction Science and Technology*, 2nd ed., CRC Press/Taylor & Francis Press, Boca Raton, FL.

9. T. S. Eyre (1978) "The mechanisms of wear," *Tribology International*, Vol. 11 (2), pp. 91–96.

10. H. C. Meng and K. C. Ludema (1995) "Wear models and predictive equations: Their form and content," *Wear*, Vol. 181–183 (2), pp. 443–457.

11. ASTM G65 (2014) "Standard test method for measuring abrasion using the dry sand/rubber wheel apparatus," Vol. 03.02, ASTM International, W. Conshohocken, PA , pp. 257–268.

12. K. G. Budinski (1988) *Surface Engineering for Wear Resistance*, Prentice Hall, Englewood Cliffs, NJ, p. 19.

13. ASTM G174-04 (2014) "Standard test method for measuring abrasion resistance of materials by abrasive loop contact," in *ASTM Annual Book of Standards*, Vol. 03.02, ASTM International, W. Conshohocken, PA, pp. 735–739.

14. ASTM G81-97a (1997) "Standard Test Method for Jaw Crusher Gouging Abrasion," ASTM Annual Book of Standards, Vol. 03.02, ASTM International, W. Conshohocken, PA.

15. P. J. Blau (1985) "The relationship between scratch and Knoop microindentation hardness and implications for the abrasive wear of metals," *Microstructural Science*, Vol. 12, p. 293.

16. M. G. Gee (2001) "Low-load multiple scratch tests of ceramics and hardmetals," *Wear*, Vol. 250 (1–12), pp. 264–281.

17. P. J. Blau, E. P. Whitenton, and A. Shapiro (1988) "Initial frictional behavior during the wear of steel, aluminum, and poly (methylmethacrylate) on abrasive papers," *Wear*, Vol. 124 (1), pp. 1–20.

18. R. T. Spurr (1965) "The wear rate of coins," *Wear*, Vol. 8 (6), pp. 487–489.

19. L. E. Samuels (1982) *Metallographic Polishing by Mechanical Methods*, 3rd ed., ASM International, Materials Park, OH, p. 34.

20. L. E. Samuels (1992) "Polishing wear," in *ASM Handbook, Vol. 18, Friction, Lubrication, and Wear Technology*, P. J. Blau, ed., ASM International, Materials Park, OH, pp. 191–197.

21. P. R. Ryason, I. Y. Chan, and J. T. Gilmore (1990) "Polish wear by soot," *Wear*, Vol. 137 (1), pp. 15–24.

22. J. Xin, W. Cai, and J. A. Tichy (2010) "A fundamental model proposed for material removal in chemical-mechanical polishing," *Wear*, Vol. 268, pp. 837–844.

23. T. K. Doi, I. D. Marinescu, and S. Kurokawa (2012) *Advances in CMP Polishing Technologies*, Elsevier, Oxford, United Kingdom.

24. F. P. Bowden and D. Tabor (1986) *Friction and Lubrication of Solids*, Clarendon Press, Oxford, United Kingdom (originally published in 1950).

25. P. J. Blau (1979) *Interrelationships among Wear, Friction, and Microstructure in the Unlubricated Sliding of Copper and Several Single-Phase Binary Copper Alloys*, PhD Dissertation, The Ohio State University, Columbus, OH.

26. J. F. Archard (1980) "Wear theory and mechanisms," in *ASME Wear Control Handbook*, M. P. Peterson and W O. Winer, eds., ASME, New York, pp. 236–241.

27. R. Holm (1958) *Electric Contacts Handbook*, Springer-Verlag, Berlin, Germany.

28. N. P. Suh (1973) "The delamination theory of wear," *Wear*, Vol. 25, pp. 111–124.

29. P. J. Blau (1981) "A simple method for cross-sectional examination of wear flakes," *Wear*, Vol. 66, pp. 257–258.

30. P. J. Blau (2003) "Microstructure and detachment mechanism of friction layers on the surface of brake shoes," *Journal of Materials Engineering and Performance*, Vol. 12 (1), pp. 56–60.

31. M. Woydt and R. Washe (2010) "The history of the Stribeck curve and ball bearing steels: The role of Adolf Martens," *Wear*, Vol. 268 (11–13), pp. 1542–1546.

32. C. M. Taylor (1993) *Engine Tribology*, Elsevier, Oxford, United Kingdom.

33. A. Kapoor, S. C. Tung, S. E. Schwartz, M. Priest, and R. S. Dwyer-Joyce (2001) "Automotive tribology," in *Modern Tribology Handbook*, Vol. 2, B. Bhushan, ed., CRC Press, Boca Raton, FL, pp. 1187–1229.

34. J. R. Davis (1992) "Friction and wear of internal combustion engine parts," in *ASM Handbook, Vol. 18, Friction, Lubrication, and Wear Technology*, P. J. Blau, ed., ASM International, Materials Park, OH, pp. 553–562.

35. R. B. Waterhouse (1972) *Fretting Corrosion*, Pergamon Press, Oxford, United Kingdom.

36. G. X. Chen and Z. R. Zhou (2001) "Study on transition between fretting and reciprocating sliding wear," *Wear*, Vol. 250, pp. 665–672.

37. R. Gresham (2015) "A little fretting about fretting," *Tribology & Lubrication Technology*, Vol. 71 (7), pp. 26–28.

38. R. D. Mindlin (1953) "Compliance of elastic bodies in contact," *ASME Transactions, Journal of Applied. Mechanics, Series E*, Vol. 16, pp. 327–344.

39. M. H. Zhu and Z. R. Zhou (2011) "On the mechanisms of various fretting wear modes," *Tribology International*, Vol. 44 (11), pp. 1378–1388.

40. R. B. Waterhouse (1992) "Fretting wear" in *ASM Handbook, Vol. 18, Friction, Lubrication, and Wear Technology*, P. J. Blau, ed., ASM International, Materials Park, OH, pp. 242–256.

41. P. A. Engle (1976) *Impact Wear of Materials*, Elsevier Science, Amsterdam, The Netherlands.

42. P. A. Engle (1992) "Impact wear," in *ASM Handbook, Vol. 18, Friction, Lubrication, and Wear Technology*, P. J. Blau, ed., ASM International, Materials Park, OH, pp. 263–270.

43. I. M. Hutchings (1992) "Wear by solid particle impact," in *Tribology—Friction and Wear of Engineering Materials*, CRC Press, Boca Raton, FL, pp. 171–197.

44. M. Welland (2009) *Sand: The Neverending Story*, University of California Press, Los Angeles, pp. 293–294.

45. ASTM G73-10 (2014) "Standard test method for liquid impingement erosion using rotation apparatus," in *ASTM Annual Book of Standards*, Vol. 03.02, ASTM International, W. Conshohocken, PA, pp. 287–304.

46. ASTM G105-02 (2014) "Standard test method for conducting wet sand/rubber wheel abrasion tests," in *ASTM Annual Book of Standards*, Vol. 03.02, ASTM International, W. Conshohocken, PA, pp. 450–458.

47. ASTM G75-07 (2014) "Standard test method for determination of slurry abrasivity (Miller number) and slurry abrasion response of materials (SAR number)," in *ASTM Annual Book of Standards*, Vol. 03.02, ASTM International, W. Conshohocken, PA, pp. 306–326.

48. F. G. Hammitt (1980) "Cavitation and liquid impact erosion," in *Wear Control Handbook*, M. B. Peterson and W. O. Winer, eds., ASME, New York, pp. 161–230.

49. C. M. Preece (1979) "Cavitation erosion," in *Treatise on Materials Science and Technology—Vol. 16—Erosion*, C. M. Preece, ed., Academic Press, New York, pp. 249–308.

50. ASTM G134-95 (2014) "Standard test method for erosion of solid materials by cavitating liquid jet," in *ASTM Annual Book of Standards*, Vol. 03.02, ASTM International, W. Conshohocken, PA, pp. 577–588.

51. H. Prashad (2001) "Appearance of craters on track surface of rolling element bearings by spark erosion," *Tribology International*, Vol. 34, pp. 39–47.

52. H. Prashad (2006) *Tribology in Electrical Environments*, Elsevier, Amsterdam, the Netherlands.

53. A. Martínez, F. Martínez, M. Velázquez, F. Silva, I. Mariscal, and J. Francis (2012) "The density and momentum distributions of 2-dimensional transonic flow in an LP-steam turbine," *Energy and Power Engineering*, Vol. 4 (5), pp. 365–371. doi:10.4236/epe.2012.45048.

54. I. V. Kragelsky, M. N. Dobychin, and V. S. Lombalov (1982) *Friction and Wear Calculation Methods*, Pergamon Press, Oxford, United Kingdom, p. 219.

55. ISO Standard 281:2007 (approved 2010) "Rolling bearings—Dynamic load ratings and rating life."

56. J. Van Rennselar (2014) "Tribological bearing testing," *Tribology & Lubrication Technology*, Vol. 70, pp. 38–46.

57. J. Halmi and P. Andersson (2009) "Rolling contact fatigue and wear fundamentals for rolling bearing diagnostics—State of the art," *Proceedings of the Institute of Mechanical Engineers, Part J, Engineering Tribology*, Vol. 224, p. 377.

58. P. J. Blau, L. R. Walker, H. Xu, R. Parten, J. Qu, and T. Geer (2010) *Wear Analysis of Wind Turbine Gearbox Bearings*, Oak Ridge National Laboratory, Technical Memo, ORNL/TM-2010/59, Oak Ridge, TN, 51 pp.

59. P. J. Blau, T. Yamamoto, K. M. Cooley, and K. S. Reeves (2013, May) "Investigation and characterization of micro-pitting in bearing steels," presented at STLE Annual Meeting, Detroit, MI.

60. M. M. Stack (2016) *Tribocorrosion for Materials Engineers*, CRC Press, Boca Raton, FL.

61. P. J. Blau, J.-P. Celis, and D. Drees, eds. (2013) *Tribo-Corrosion: Research, Testing, and Applications*, ASTM International, W. Conshohocken, PA (ASTM STP 1563).

62. P. Lambrechts, E. Debels, K. Van Landuyt, M. Peumans, and B. Van Meerbeek (2006) How to simulate wear? Overview of existing methods," *Dental Materials*, Vol. 22, pp. 693–701.

63. P. J. Blau (1989) *Friction and Wear Transitions of Materials: Break-in, Run-in, Wear-in*, Noyes Publishing, Park Ridge, NJ.

64. C. Pliny (2015) "Neanderthal teeth suggest sexual division of labor," http://www .redorbit.com/news/science/1113338174/neanderthal-teeth-suggest-sexual-division -of-labor-022015.

65. P. J. Blau (2015) "How steady is the steady-state? The implications of wear transitions for materials selection and design," *Wear*, Vols. 330–331, pp. 1120–1128.

66. J. C. Miranda and A. Ramalho (2015) "Study of the effects of damage accumulation on wear," *Wear*, Vols. 330–331, pp. 79–84.

Chapter 4

Tools for Imaging and Characterizing Worn Surfaces

Chapter 3 provided a means to identify the forms and characteristics of wear and surface damage based on surface appearance and the contributory processes. Visual examination and enhanced forms of imaging have become the key tools for identifying and diagnosing wear in the field and in the laboratory. This chapter presents some common and lesser-known tools for wear surface characterization and some of the related techniques with particular utility for wear mode diagnosis. By far, optical examination by naked eye and light optical microscopy (LOM) are the most common. Since the mid-1900s, when it was introduced and then spread quickly into the research community, electron microscopy—particularly, scanning electron microscopy (SEM)—has become an almost routine tool for wear diagnosis. LOM was less expensive and did not require the kinds of special specimen preparation procedures that SEM did. Therefore, while SEM provided very high magnification, three-dimensional (3D) imaging options, and the kinds of large depths of field needed for rough wear surfaces, the specimen sizes and kinds of materials suitable for examination by today's SEM are still more limited than for LOM.

Having detected wear using the methods in Chapter 2, and after identifying the operating form(s) of wear using the criteria in Chapter 3, the next step is to apply appropriate visual examination methods to home in on the details. Sometimes, the naked eye or a ×10–×30 magnifying loupe is all that is needed to shed light on the problem, but there are other times when more sophisticated tools and techniques, like those described later in this chapter, are needed. While research quality instruments can be helpful in focusing on the details, a lot can be learned from simpler and more readily available methods of examination and optical metallography.

While rarely used these days in formal reports, it is often advantageous to use hand sketches in field notes to document wear damage. One advantage of a sketch is that our eyes can sometimes discern features that are difficult to capture by photography (due to glare or lack of contrast). Drawings can be highlighted with arrows and other annotations simply and directly. Note, for example, Figure 4.1, which is a series of three sketches from a technical report on wind turbine gearbox bearings [1]. The key features of these worn components can easily be indicated in this way and serve as a map when more detailed, enlarged images of specific areas are later obtained by microscopy. Arrows can be added to the sketches to show places where high magnifications were used in associated images.

After making an initial naked-eye examination, there are a wide variety of light and electron optical imaging options for wear diagnosis (see Figure 4.2). The cost and sophistication level of these tools tend to increase from left to right, although technically, the human eye is the most sophisticated imaging system of all. Advancements in the field of imaging continue to add enhanced features like multiple image stitching and auto focusing.

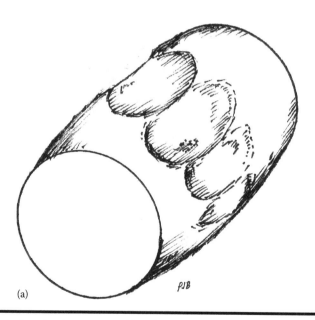

(a)

Figure 4.1 **A sketch of severe spalling fatigue damage on an overloaded spherical bearing rolling element from a wind turbine gearbox after service at a wind farm in northern Colorado. (a) Equatorial spalling and smearing.** *(Continued)*

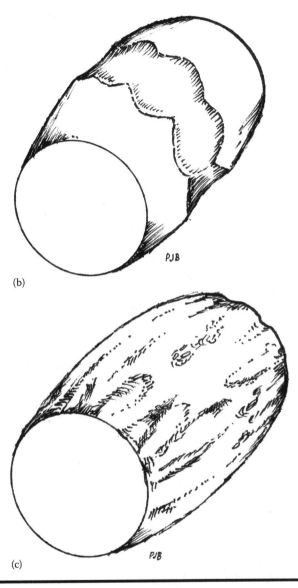

(b)

(c)

Figure 4.1 (Continued) A sketch of severe surface damage on two spherical bearing elements from a wind turbine gearbox after service at a wind farm in northern Colorado. (b) Connected equatorial plastic deformation. (c) Spalling, flaking, and grooving on the entire crowned surface. (From Blau, P.J., Walker, L.R., Xu, H., Parten, R., Qu, J., and Geer, T. *Wear Analysis of Wind Turbine Gearbox Bearings,* ORNL Technical Report, 2010 [1].)

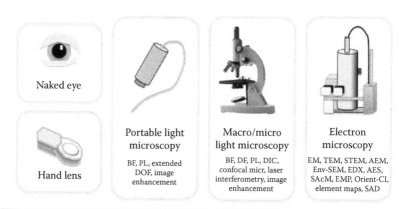

Naked eye	Portable light microscopy	Macro/micro light microscopy	Electron microscopy
	BF, PL, extended DOF, image enhancement	BF, DF, PL, DIC, confocal micr, laser interferometry, image enhancement	EM, TEM, STEM, AEM, Env-SEM, EDX, AES, SAcM, EMP, Orient-CI, element maps, SAD
Hand lens			

Figure 4.2 Light and electron optical imaging choices. (BF = bright field, PL = polarized light, DOF = depth of field, DF = dark field, DIC = differential interference contrast, EM = electron microscopy in general, TEM = transmission electron microscopy, STEM = scanning transmission electron microscopy, AEM = analytical electron microscopy, Env-SEM = environmental scanning electron microscopy, EDX = energy dispersive X-ray spectroscopy, AES = Auger electron spectroscopy, SAcM = scanning acoustic microscopy, EMP = electron microprobe analysis, Orient-CI = orientation contrast imaging, SAD = selected area diffraction.)

There are two complementary methods of wear surface examination: normal surface viewing and cross-sectional viewing. Normal viewing refers to viewing perpendicular to the as-worn surface. It is used at relatively low magnification (say ×5–×50) to get a general idea of the morphology before zooming into the finer details. Cross-sectioning of wear surfaces requires more time, effort, and special metallographic techniques, but if its use is warranted, it can provide valuable information to help diagnose wear types, identify directions of relative motion, determine the extent of microfracture, and ascertain the depth of contact damage. Light optical microscopy will be described first, along with methods for cross-sectioning, including the helpful technique of taper-sectioning.

4.1 Light Optical Microscopy

The term *light optical microscopy* (LOM) is preferred to the common term *optical microscopy* because other instruments such as the transmission and scanning electron microscopes discussed later in Chapter 4 also use optical systems—but they use electron optics, not visible light optics. LOM covers a wide range of magnifications, ranging from a magnifying glass (×2–×5) or jeweler's loupe magnifying from ×10 to ×60 to a high-resolution light optics with software that can exceed magnifications of ×1000.

There are numerous textbooks available on the subject of LOM. They describe the basic principles of imaging theory, focal length, depth of field, image aberrations,

lens combinations, diaphragm adjustments, alignment, lens coatings, and techniques used for both transmitted and reflected light illumination. This book makes no attempt to address the basic principles of LOM but rather highlights imaging modes and methods that can be useful for wear diagnosis.

A selection of magnifiers is shown in Figures 4.3 and 4.4, which shows a portable digital microscope with polarized light and zoom capabilities (about ×20 to ×200) that in this case is powered by the computer's Universal Serial Bus (USB) port. Fortunately, handheld digital microscopes, like that shown in Figure 4.4, are readily available to aid in field work. Hitachi Instrument Company has built SEMs capable of imaging objects that are about 1 m³ in size, but such facilities are rare and expensive to maintain. Even if such an instrument were available, the need to operate under high vacuum limits their use on "dirty," "greasy," or crevice-filled specimens such as large bearings and gearsets.

Various image processing options are available with digital microscopes. For example, these can compensate for glare on metallic specimens and provide a large apparent depth of field using what is called *High Dynamic Range* (HDR) and *Extended Depth of Focus* processing, respectively. Both methods involve integrating several images over a range of incremental optical settings. In HDR, a series of images is acquired automatically at different effective brightness levels, and the best portions of the image at each exposure level are integrated into the final image. For example, Figure 4.5a and 4.5b compare two images of sand particles in which HDR has improved the detail in both the bright and darker portions of the image.

Figure 4.3 The author's collection of magnifiers and loupes. The large lens has a ground spot near the handle that magnifies more than the larger portion, and the jeweler-style loupes (upper right) have built-in battery-powered LED illuminators and dual magnifications.

Figure 4.4   **Portable digital microscopes such as this Dinolite™ instrument offer compact size and portability. Mounted on a stand, it is about 10 cm long, and is powered by a USB port from a laptop computer. The wider ring, halfway up the barrel, controls magnification. The ring at the bottom adjusts the polarizer.**

4.1.1 Specimen Mounting

The mounting of large worn specimens for magnified viewing can be a challenge due to matters of shape, size, and weight. It is often desirable to examine a component in the field, and for some large components, it may be necessary to bring the microscope, such as the portable unit in Figure 4.4, to the specimen rather than attempt to bring the specimen into the laboratory.

Replicating compounds. When a worn part cannot be easily cut out or removed from a machine for microscopic examination, replicating compounds may be used. Wear features from a large part can be replicated with a specially formulated, castable compound that provides submicrometer resolution of surface features under ideal conditions then removed and mounted for microscopy. One such family of surface replicants is known as Microset™. It uses a silicone rubber compound that can be cast on a surface, removed, and provides submicrometer feature resolution [2]. Bear in mind that the worn areas may need to be carefully cleaned of lubricant

(a)

(b)

Figure 4.5   Images of mixed sand grains taken with the portable digital microscope shown in Figure 4.4 under normal illumination (a), and using HDR postprocessing software (b) to level the brightness and bring out more details. The oblong particle in the center is about 2.3 mm long.

residues or debris in order for the compounds to work correctly. Therefore, lubricated surfaces or scaly, corroded surfaces may be unsuitable unless considerable preapplication care is taken to clean them. This of course creates the problem of disturbing or removing potential clues such as debris, contaminants, and tribofilms during subsequent cleaning and mounting.

4.1.2 Specimen Cleaning

The cleaning of retrieved wear surfaces can potentially remove some clues to the type of damage, so it is useful to examine as-retrieved wear parts before and after cleaning. A careful inspection of Figure 4.6a shows metallic debris flakes within a grease deposit surrounding a reciprocating, crossed-cylinders test. Wear testing using greases is a challenge because the grease moves around and must be supplied consistently to the interface. Figure 4.6b is from a similar test in that series, but after the grease was removed to reveal the wear details. Together, the two images present a more complete picture of the wear process than either by itself.

The manner of cleaning specimens for wear analysis is a large subject in itself because the method chosen must depend upon the materials and the physical nature of the worn samples and their environment. For example, nonlubricated wear parts can be lightly brushed off or dusted off with a compressed air jet (canned air or clean bottled air is better than centrally piped air because the latter often contains oil from the pumps). Oily parts from engines can be rinsed off with gasoline, followed by alcohol or acetone then methyl alcohol. Remember that once the oil is removed, metallic surfaces can begin to rust or corrode, so preservation of the wear surface may require reoiling, wrapping in oiled paper, or storing in a controlled environment to retard complications from postwear corrosion. For some types of plastics, cleaning in acetone may not be a good idea, so mild soapy water with a thorough distilled water rinse and dry might work better. Part of the problem with cleaning, especially if weight loss is used to determine wear, is that one needs to remove any adherent deposits that can mask the true weight loss yet in doing so not abrade or otherwise disturb the wear features if microscopy is to be used on them later. Some trial-and-error experimentation with cleaning methods on less critical parts of the piece is worthwhile.

Cleaning may be necessary during wear testing in order to measure wear by weight change or profiling. Again, one should inspect the specimens for clues such as debris residues and optionally remove samples of such materials before cleaning. Depending on the type of specimen, the sensitivity of the materials, and wear environment, the following cleaning methods can prove helpful:

- Brushing with a stiff bristled brush (avoid wire brushes, especially if microscopy is to be done later)
- Dusting off with compressed air (some central air systems can contain oil from the compressors, so purer bottled air may work better)
- Water jet or solvent jet spraying (squeeze bottle)
- Wiping with solvent-soaked cloth or no-lint laboratory wipes
- Ultrasonic cleaning ("sonication") using a series of solvents (gasoline, acetone, ethanol)
- Vapor degreasing
- Vacuum oven "bake-out"
- Ultraviolet (ozone) degreasing such as that used in the semiconductor industry

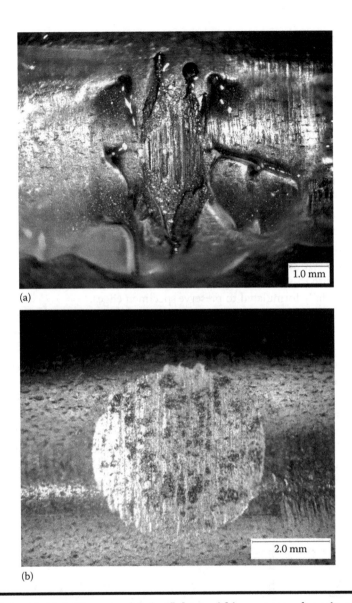

(a)

(b)

Figure 4.6 (a) Metallic wear debris (flakes) within a grease deposit on a wear scar from a reciprocating test indicates the presence of adhesive wear. (b) A steel specimen from the same set of tests, but after the grease was removed to reveal features associated with adhesive wear and minor third-body abrasion from loose, work-hardened debris. (LOM images using a macroscope equipped with a digital imaging head and calibrated scale bar software.)

4.1.3 Cross-Sectioning

The mounting of large worn specimens can be a challenge for microscopic examination. For some large components, it may be necessary to bring the microscope to the specimen instead of the opposite. Fortunately, handheld digital microscopes are readily available. These compact units can be powered by a portable computer and offer special features like extended dynamic range (to reduce glare and reveal features in shadow), polarized light (for glare control), long depth of field to image rough areas, and integrated image stitching that facilitate documentation.

Cross-sectioning of worn surfaces provides valuable information on the depth of damage and type of wear. However, since the cutting and polishing process that is used to create cross-sections can also damage or obscure subtler wear details, special methods for edge preservation are sometimes needed in the mounting stage. These include electroless Ni plating prior to sectioning (check which metals are compatible with this method) and the use of hard resin-based mounting media that has been specially formulated to preserve specimen edges.

Essentially, a cross-section represents the intersection of three damaged zones (see Figure 4.7). Below the original manufactured surface (dotted line O–C), one finds a region of nominal depth C–D that contains artifacts left from forming and finishing the load-bearing surface. The depth of apparent wear damage on the cross-section viewed from the left does not necessarily equal the total amount of material removed from the original surface dashed line (O–C). Therefore, wear depth measurements made without a reference point to establish the original surface are those remaining after the above material has been removed. Also shown schematically in Figure 4.7 is a thin zone of microstructural alteration from the process(es) used to cut and polish the cross-section in the plane of the cross-section that derives from the methods used to create the cross-section. The challenge is

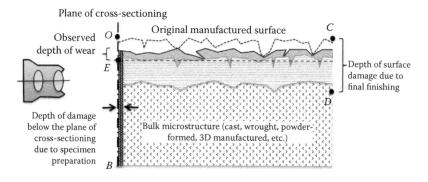

Figure 4.7 **Intersection of damaged zones when a machined and finished metal specimen has been worn and is then cross-sectioned for near-surface examination.**

to recognize which features come from which of the possible sources. Sometimes, continuing wear can remove surface finishing artifacts entirely, but sometimes, the wear is shallow enough to remove only a portion of the as-machined surface damage. In general, measured roughness values on engineering surfaces are not indicators of the depth of residual machining and finishing damage. It is possible to have a smooth, polished-looking surface that retains a highly textured, work-hardened layer. Consider, for example, the lubricated wear of cylinder bores. A wear process might simply remove the tips of the tallest asperities but not affect the as-ground material that surrounds deeper finishing grooves.

Specialized edge-preservation procedures, like electroless plating before cross-sectioning, the use of thin diamond wafer saws, and others described by Samuels [3], can be used to minimize specimen preparation artifacts. Edge rounding is a problem during polishing specimens like softer metals because the area of greatest interest, the near-surface layers in the specimen, can be rounded if the mounting material adjacent to it is too soft to protect it. Edge rounding causes problems when examining the edge zones with high LOM magnification that tends to have shallow depth of focus. A related problem is the loss of loosely adherent wear debris particles during polishing. This is also why it is best to examine cross-sections as-polished first and before the application of any etchants that could attack near-surface debris deposits and remove clues for wear process diagnosis.

In his classic book, Samuels [3] has compiled data on the depth of damage produced by various machining and finishing methods. Therefore, to remove artifacts from each step of specimen preparation, the next stage should remove at least the amount of material damaged by the previous grinding or polishing step.

4.1.4 Taper-Sectioning

While much can be revealed by cross-sectioning a worn surface perpendicular to the sliding surface, a variation of this method called *taper-sectioning* deserves a special note. It produces an artificial magnification of surface features by tilting the plane of polish at a shallow angle to the surface. The magnification effect occurs in the plane of the taper but not transverse to it. This is analogous to a stylus profile in which the vertical and horizontal magnifications are unequal. The effective magnification M of elongated features is equal to the cosecant of the angle of incline (θ in Figure 4.8).

The basic taper-sectioning procedure is as follows:

1. Clean the wear surface and, if possible, apply a protective plating to preserve the wear scar. Not all kinds of materials can be electroplated.
2. Prepare a platform (wedge) with the desired taper angle.
3. Using a drop of epoxy, mount the specimen of interest on the wedge with the wear surface lying across the wedge and the direction of desired magnification in the direction of tilt. Apply a release agent such as a sprayed Teflon

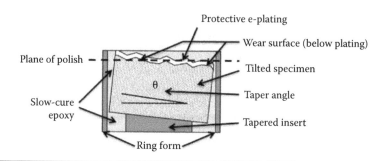

Figure 4.8   Principle of taper-sectioning. In this case, an electroplated wear specimen is mounted on an insert placed inside a ring and then cast in metallographic epoxy.

(PTFE) film or silicon grease to the plate glass to support the mount while the mounting medium is setting.
4. Cast the specimen and its wedge in a high-grade metallographic, slow-curing epoxy. Ensure that the top layer of the epoxy is free from gas bubbles when gently pouring it into the mold.
5. Polish the specimen as shown until the edge of interest is exposed.
6. Examine by LOM. Then etch if required, working gradually to avoid over-etching the near-surface region.

A taper-sectioned wear surface of brass is shown in Figure 4.9. Note that any measurements made on such specimens must be compensated for the taper magnification and direction. The distance across the wedge is the original microscope image magnification, but the size of features that lie in the direction of the tilt are magnified by M (= cscθ).

4.1.5 Transmitted versus Reflected Light Illumination

The two basic modes of LOM are based on the direction of specimen illumination and degree of transparency. They are transmitted light and reflected light. By far, reflected light (normal incidence) is used for examining wear specimens. Transmitted light is commonly used mainly for biological specimens, for examining thin petro-sections in mineralogy, or for studying transparent materials in general. The balance of this discussion will therefore concern reflected light methods.

The type of illumination, its wavelength, and its direction/angle of illumination to the surface, coupled with the instrument design, affects the brightness, contrast, resolution, and visibility of wear features. Some methods work better for shiny metallic surfaces and others work better for matte surfaces or surfaces that diffuse the light. For example, abraded ceramic tiles can be difficult to examine by reflected light of normal incidence. Tilting them to increase surface reflections and

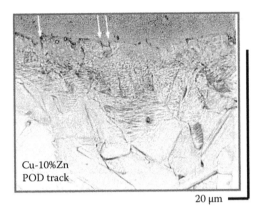

Cu-10%Zn
POD track

20 μm

Taper magnification = cscθ

Figure 4.9 Taper-sectioned specimen of Cu-10%Zn slid against Type 52100 hardened steel under dry conditions. After Cu plating, taper mounting, and polishing, the specimen was etched to reveal the grain-to-grain response to sliding contact. Arrows indicate microcracks at the edges of remaining plateaus.

show scratches can be useful, but a compromise must then be made when tilting because the depth of focus tends to be limited for normal compound microscopes. Sometimes, sputter coating surfaces with Au-Pd and using SEM provide better imaging contrast for matte surfaces. Therefore, SEM is used not only for high magnifications but also to provide a different form of contrast enhancement because the mechanism of developing contrast (secondary electrons versus visible light photons) is not the same.

Angle of illumination, glare, and light diffusion. When illuminating wear surfaces, it is worthwhile to experiment with the intensity and direction of illumination in order to bring out the features of most interest. Imaging bearing balls, rollers, and metallic shafts is particularly challenging due to glare and reflections that can obscure wear features. Simple strategies like the use of white reflectors (a rolled up piece of paper or a ring light around the microscope lens, for example) can be effective. Other approaches are to use optical diffusers or multiple light sources. The author sometimes uses a flexible optical light pipe illuminator to direct a beam at angles oblique to the surface. This can be effective even on a traditional compound microscope by turning off the microscope illuminator and using a light pipe aimed along the surface instead.

Monochromatic versus polychromatic illumination. Polychromatic (white) light illumination is usually used with hand lenses, loupes, and macroscopes. If the imaging device is not designed to control chromatic aberration, higher magnification of bright metallic objects may produce annoying color fringes around the features or edges. To avoid such artifacts, metallurgical grade microscopes tend to use monochromatic (green) illumination. In that way, sharper focus, crisper images, and higher feature resolutions may be achieved. Obviously, the ability to identify features by color is lost when using monochromatic light.

There is an excellent discussion of the principles of illumination and optical system components in Chapter 4 of the text by Vander Voort [4]. Table 4.1 summarizes several LOM examination methods. There are a number of other methods available, including infrared, ultraviolet, and fluorescence microscopy, but these tend to find primary application in biological sciences and fields other than tribology. When considering the options in Table 4.1, note that each technique has limitations and that the method of producing contrast and color is not the same among the methods. Some techniques, such as interferometry, do not work as well on very rough surfaces with overhangs of material or surfaces that are not highly reflective.

4.2 Stylus Profiling and Noncontact Topographic Imaging

Wear is a transition between the unworn and the worn condition, and there is a common intuition that wear makes surfaces rougher. That is often not the case. Consider running-in of ground metal surfaces in which early sliding contact truncates the peaks of asperities and leads to a smoother steady-state condition that is better for lubrication and performance in general. In other cases, smooth starting surfaces roughen due to plastic deformation, plowing, scoring, fracture, pitting, and other phenomena. Even scuffing, a phenomenon often associated with a rise in surface damage, can actually make a surface smoother rather than roughening it.

The profiling of surfaces has been enabled by the development of stylus tracing instruments and noncontact optical instruments. Like any tool, each method has advantages and limitations, and the reader is referred to published reviews in that area (e.g., Song and Vorburger [5] and Bhushan [6]). The primary uses of such methods in wear analysis are as follows:

1. To verify that the roughness of engineered surfaces meets design specifications
2. To quantify changes in surface feature dimensions (standard roughness parameters) due to wear
3. To provide visual or numerical information concerning the directionality or heterogeneity of worn surfaces
4. To measure the volume of material removed or displaced by wear

The subject of surface texture (which includes aspects of directionality ["lay"], waviness, and roughness) is a broad specialty field, applied to both manufacturing technology and scientific studies of the functionality of surfaces, be it electro-optical, mechanical, cosmetic, or tribological. In wear analysis, we focus on the change in morphology, but the kinds of rough worn surfaces being examined can sometimes make the use of some surface imaging methods difficult or impractical.

Table 4.1 Common Illumination Options in Light Optical Microscopy of Wear Surfaces

Method	Principle	Use
Bright-field (polychromatic or monochromatic)	Light transmitted through a transparent/translucent specimen or reflected back from the surface	Most common illumination method in LOM; used for metallographic specimens
Oblique illumination	Can be done by manipulating the microscope condenser alignment or by changing the position of the light source	Can enhance wear features and enhance the three-dimensional appearance of features like grooves or lumps
Dark-field illumination	Placing a central blockage in the light path and opening the apertures such that only light bouncing back at an angle from the surface appears in the image	Can accentuate wear features but tends not to be as sharp and clear as bright-field; helpful to detect scratches while causing a normally bright, flat background to appear dark
Polarized light	Use of one or two polarizers in the illumination and/or postreflection optical train; image contrast, color, and brightness adjusted by rotation of one or both polarizing filters	Helpful in glare reduction from shiny areas, enhancing image colors, identifying minerals, seeing below the surfaces of semitransparent layers such as oxide films, some ceramics, or some kinds of abrasives
Sensitive tint (a.k.a., red tint plate)	Uses a slice of birefringent material to add color contrast to polarized images; suppresses green light and emphasizes reddish tones	Can add colors to microstructural features to improve grain–grain contrast in cross-sectioned polyphase materials, for example

(Continued)

Table 4.1 (Continued)   Common Illumination Options in Light Optical Microscopy of Wear Surfaces

Method	Principle	Use
Phase contrast (a.k.a., differential interference contrast [DIC] and Nomarksi interference contrast)	Uses an extra prism in the objective lens that can be rotated; works by combining interferometric principles with polarized light	Can improve contrast and differentiate fine features based on height differences of only a few nanometers; can be useful in bringing out the deformation or twinning features in cross-sectioned highly deformed layers that have been lightly etched
Confocal microscopy	Any of several variations of optical methods, some involving lasers, that can build images by varying the plane or focus sequentially and integrating features of sharpest focus into a single image	Can be used to map height differences in images, measure features, and improve depth of sharp focus in images; can be useful in wear surface imaging and measurements but has limitations for very rough surfaces
Vertical scanning interferometery	Uses automated focusing steps with a reference beam and a beam reflected from the surface to produce contrast from small height differences	Software can produce topographic or pseudo-3D images that can be used for measurements of surface features or computing surface finish parameters (see also Section 4.2)

For example, optical interferometry generally involves "counting fringes" on images and wear surfaces with sharp, steep features, and overhanging materials, or even clumps of debris particles, which can make portions of the image unmeasurable. Contact stylus profiling is limited by the shape and radius of the stylus tip, the directionality of the features it is traversing, the deformation of soft materials under the action of the diamond tip, and even the frequency response of the recording instrument (the faster the tip moves, the more quickly any localized changes in topography must be detected).

Figure 4.10 illustrates a topographic image of galled stainless steel obtained using a commercial vertical scanning interferometer and colored to show the height

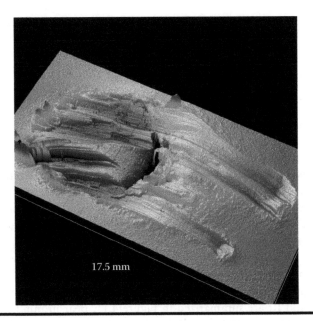

Figure 4.10 **3D image of a galled area on a stainless steel specimen obtained using a commercial vertical scanning interferometer. Height-coded shading emphasizes the topography and the characteristic excrescences that accompany galling damage.**

from a reference plane. The software in such systems also enables feature measurement. The wear depth or volume of material removed or displaced can be obtained using various subroutines in the software.

The 3D rendering of surfaces offers a powerful advantage over a two-dimensional (X-Z) mechanical stylus trace taken at a specific location and traverse direction. Generally, the height dimension is magnified relative to the length dimension in stylus instrument plots so the asperities look much sharper and taller than they would if both the X and Z scales were the same.

Equipment has been developed to do a raster type of stylus scan by stacking up individual traces taken at some finite distance apart and stitched to produce a quasi-3D image. Such instruments are less popular than noncontact methods because they require much longer to generate an image and tend to be costly. They are not subject to reflectivity issues (see Section 4.2.1) but suffer from the limitations of finite tip radius, tip wear, frequency response, and the fact that there are no data between adjacent traverses, so some features can be missed. Stylus profilometry data can also be affected by surface lay (the directionality of features). Traversing at some angle to wear grooves can tend to pull the stylus to the side unless it is very stiff in every direction except the vertical direction. Since surface texture consists of roughness, waviness, and lay, the user must decide on the traverse direction of a contact-type device relative to the lay. Conventionally, but not always, the roughness of a wear or engineering surface is measured perpendicular to the lay.

4.2.1 Comparison of Roughness Data from Different Measurement Methods

The measurement of surface roughness by stylus methods and noncontact methods by optical systems has certain pitfalls. There are inherent resolution limitations with each technique, and the characteristics of the surface (deep cracks, highly tilted walls, reentrant features like overhangs, dull finishes, poor reflectivity, software filters, and more) can affect surface roughness parameters. In a previous study, surface roughness parameters using four methods of measurement were compared on a reflective metal and a lower-reflectability ceramic [7]. The materials are listed in Table 4.2, and the methods used to make the surface finish measurements are shown in Table 4.3. Three standard surface roughness parameters were measured: (i) arithmetic average roughness (Ra), root-mean-square roughness (Rq), and peak-to-valley roughness (Rt). Since Ra is similar in magnitude to Rq, the results of this study for Ra and Rt are summarized in Figure 4.11. Bear in mind that only two specimens were used and that four different measurements were used on each one.

The important messages to be gained from the foregoing study are the following:

1. When specifying a certain surface finish for a product, the supplier and the customer should agree on the type of measurement equipment to be used (same type and model of equipment if possible or else agree on shared calibration standards).

Table 4.2   Materials Used in a Surface Measurement Methodology Study

Surface Characteristic	Material Description	Method of Preparation
Matte finish	Silicon nitride, polycrystalline ceramic tile (commercial designation NCX 5102, Norton-St. Gobain, Worcester, Massachusetts)	Surface ground with a 400-grit diamond wheel, depth of cut of 127 μm, and table speed 0.127 of m/s
Polished reflective finish	Inconel alloy 625 tile similar size to the ceramic tile	Hand-ground using 400-grit wet SiC polishing papers

Source: Blau, P.J., Martin, R.L., and Riester, L., *A Comparison of Several Surface Finish Measurement Methods as Applied to Ground Ceramic and Metals Surfaces,* ORNL Technical Report, ORNL/M-4924, 1996 [7].

Table 4.3   Methods Used in a Surface Measurement Methodology Study

Type of Method	*Figure 4.11 Label*	*Measuring Conditions*
Mechanical, skidless diamond stylus (Talysurf 10™)	Stylus	Six traces at each of four locations; ISO filtered, seven cutoffs 0.25 mm long; 2 µm tip radius
IR laser scanning measuring system (Rodenstock)	Laser	12 area scans; 0.25 mm cutoff filter, 4 µm tip simulation; 20 mm/s scan speed
Atomic force microscope (AFM) (Topometrix)	AFM, AFMx	150 × 150 µm (ceramic), 100 × 100 µm (metal) scan areas; fine tip simulation (AFMx) and normal tip simulation (AFM)

Source: Blau, P.J., Martin, R.L., and Riester, L., *A Comparison of Several Surface Finish Measurement Methods as Applied to Ground Ceramic and Metals Surfaces*, ORNL Technical Report, ORNL/M-4924, 1996 [7].

2. The degree of agreement between one roughness measuring method and another depends not only the method used but also on the reflectivity and other characteristics of the materials involved.
3. It is very difficult to determine the "true surface roughness" of an engineered surface because the methodology used can produce significantly different results.

In regard to item 3, it is, in principle, possible to prepare metallographic cross-sections of the test surface in several places, digitize the resulting cross-sectional profiles at high resolution, and then compute the roughness parameters from those data. However, a lot of samples would have to be prepared and measured very carefully, making such a proposal impractical. Furthermore, the resulting values would be unlikely to correlate with measurements using a stylus or noncontact method. Therefore, the message is if surface roughness measurements are to be compared in different organizations, the same methodology should be used. If relative differences in roughness are important, as they are in some wear studies, then consistent metrology practice with roughness calibration standards becomes very important.

4.2.2 Confocal Microscopy

Developed in the 1950s, confocal microscopy is another 3D imaging and feature mapping method that has grown in use and popularity, especially in biological research. With its roots in light optical illumination, in the late 1960s, laser versions have emerged. There are a number of variants of the method, but they all basically operate using a stack of images taken at sequential planes of focus with software to

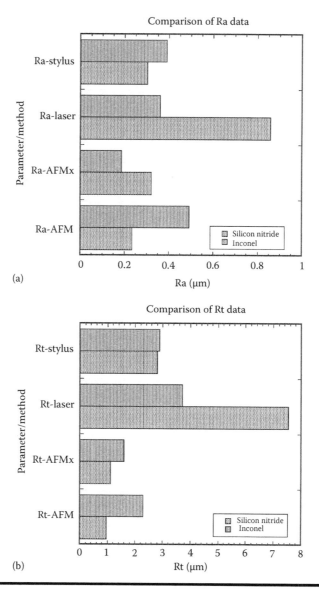

Figure 4.11 **(a) Comparison of surface roughness parameter Ra for a metal and ceramic specimen using four measurement methods. (b) Comparison of surface roughness parameter Rt for a metal and ceramic specimen using four measurement methods.**

detect the sharpest features in each image and integrate them to construct a deep-focus, fully sharpened result. With their pseudo-color renderings, confocal images look similar to those obtained by vertical scanning interferometry. Like any optical technique, it has limitations for materials having high roughness and asperity slopes, surface curvature, reentrant features (overhangs), or low reflectivity.

4.3 Scanning and Transmission Electron Microscopy

The popularity and availability of electron microscopes of various kinds have made them nearly a routine choice for wear studies. Numerous reviews of the operation and principles of transmission electron microscopes, scanning transmission electron microscopes, analytical electron microscopes, and standard and environmental scanning electron microscopes are available in books, reviews, and on the Internet. Their power lies in their ability not only to image details with high clarity at high magnification but also to obtain compositional information by energy dispersive x-ray analysis, Auger electron spectroscopy, and other chemical analysis accessories now available on electron microscopes. Recent advancements in telecommunications have made possible the ability for a user to send a specimen to a facility thousands of miles away and remotely operate an electron optical instrument at his or her desk. For example, in October 2007, researchers at Imperial College (London, United Kingdom) operated in real time a state-of-the-art electron microscope at Oak Ridge National Laboratory (near Knoxville, Tennessee) nearly 4100 miles (6584 km) distant.

In terms of theory and applications, Goldstein et al. [8] is a good general reference to imaging and compositional analysis using a SEM. Computerization has made significant advances in electron optical alignment, automated astigmatism corrections, faster pumping times, compositional accuracy, more forgiving atmospheres around the specimen (in terms of degassing of surface deposits and even living tissue), and display options. Electron imaging technology is being improved rapidly, so the reader is advised to seek current information on the subject from equipment makers and users.

A final word of caution, from a wear diagnosis point of view, the ready accessibility of SEMs and other sophisticated imaging instruments does not negate or replace the value of LOM. Therefore, the first choice in examining surfaces should be LOM or at least a hand loupe. In this way, areas of interest for SEM can be found and can save having to scan large areas to find details of interest. Furthermore, wear may differ at different locations even on the same part, so having a low-magnification overview of a larger area (in addition to notes, images, maps, and sketches) will place any unusual features into a proper perspective. The statistics of surfaces offers a powerful advantage over a two-dimensional (X-Z) mechanical stylus trace taken at a specific location and traverse direction. of beginning at low magnification is also convincing. Say, a wear scar on a bearing surface is 10 × 20 mm in size and is being examined in a SEM at moderate magnification such that the image shows an area of, say, 100 × 150 μm. That image shows only 0.0075% of the entire wear scar! One would need to look at an enormous number of such images to get a complete picture of the dominant form(s) of wear. In fact, misdiagnosis is also possible if one happens to base a full wear diagnosis on viewing only a few SEM images. LOM with selected SEM of details of interest is the prudent way to go in this case.

4.4 Nontraditional Imaging Methods

As technology marches on, new surface imaging tools inevitably become available. At least in the early stages of their development, they tend to be expensive and challenging to use, often requiring a specially trained and dedicated operator. Therefore, they are commonly used first in research. If they meet with market success, additional development may enable more "turn-key" types of operation suitable for routine use. However, as with any tool, there is a right and a wrong way to use it. For example, acoustic images look like light microscope images, but their source of contrast is entirely different (based on acoustic impendence), possibly leading to misunderstanding and misinterpretation of the observed features. Therefore, training and experience are needed to make proper use of the methodology.

Over the years, the author has used a variety of imaging tools other than those described earlier in this chapter. Some have limited application or require the specimens to be prepared in ways that make it difficult to preserve wear features. Some require extremely flat specimens to avoid noisy images or signal dropout. Others use liquid media that prevent their use on materials that react to liquids or are so porous that liquids can penetrate and produce misleading images.

4.4.1 Scanning Acoustic Microscopy

Scanning acoustic microscopy (SAcM) relies on changes in the speed of sound in materials or across discontinuities at or below the surfaces of materials to create a contrast in computer processed images. When properly applied, the technique can produce images of subsurface features, like cracks or inclusions, that are invisible to optical or electron microscopes. However, the application of the method is limited to specimens with relatively flat surfaces and those which are tolerant to water or water-glycol mixtures, the media used to couple the acoustic signal to the specimen through a specially ground sapphire lens. Figure 4.12 illustrates the components of an SAcM.

Based on high-frequency sound waves (on the order of 100s of kilohertz to a few gigahertz), the magnification of the method is limited to a few hundred times. Resolutions of less than 1 μm to tens of micrometers are typical, but the technology continues to improve. Also, owing to their acoustic impedance, some materials are

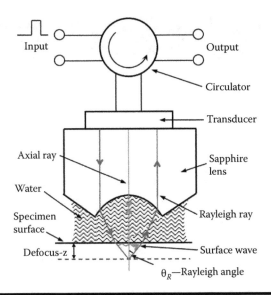

Figure 4.12   Principle of the scanning acoustic microscope.

capable of sharper and clearer images than others. While the images look like normal microscopic images, the sources of contrast are much different, and it requires training and experience to interpret them. The sound waves can be focused (or as it is sometimes called, "defocused") at different depths within the material to detect flaws (like delaminations or cracks) that reside at those depths. The higher the frequency, the higher in principle is the resolution and the lower is the depth penetration. High frequencies also make the image sensitive to artifacts and roughness. Therefore, the selection of acoustic imaging parameters is a compromise between instrument settings, the material, the roughness of the surface, and the need for deeper penetration or surface feature images.

Figure 4.13a and 4.13b compare a light optical image of an etched cross-section of a carburized stainless steel surface with that for the same specimen using SAcM. Note both the similarities and differences on contrast. The nearly featureless carburized layer in the light optical image is shown to be composed of grains in the SAcM image to its right. The contrast in the layer is due to an anisoptropy in the acoustic impedance of the crystal grains.

One of the early uses of SAcM was in inspecting for flaws in the production of electronic semiconductor chips. The flat uniform surfaces of those chips and the ability to couple the lens acoustically to the surface using a droplet of water worked pretty well, but those instruments were highly sophisticated and finicky and cost as much as, if not more than, a good scanning electron microscope. Despite this, SAcM has been applied to surface engineering research under limited conditions and has provided subsurface data that would be difficult to obtain in other ways [9,10].

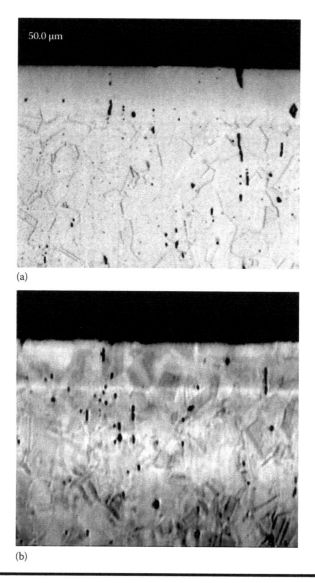

(a)

(b)

Figure 4.13 A comparison of a light optical image with an SAcM image of the same 316 stainless steel specimen treated with a proprietary low temperature, high carbon super-saturation heat treatment. (a) Light optical image. (b) Scanning acoustic image (1.0 GHz).

While useful in some aspects of tribology and machining research, the roughness of typical wear specimens that interacts with the sound waves used, the curvature of bearing surfaces, and the sensitivity to water of steels and other likely tribological specimens have largely limited the application of SAcM to tribology research. Its use in wear diagnosis is therefore limited by material suitability, instrumentation cost, artifacts on rough wear surfaces, the need for a coupling fluid, and user expertise.

4.4.2 Thermography and Thermal Wave Imaging

Thermography is based on the emission of infrared radiation. It has become an indispensible tool in many fields, including medical diagnosis, military surveillance, home insulation evaluation, fluid channel blockage, leaking pipes, industrial equipment diagnostics, and many more. Infrared thermometers are now readily available and relatively inexpensive, but like any method, they must be used with appropriate understanding of their advantages and disadvantages. To achieve accurate thermal maps, the emissivity of the surfaces being studied needs to be well known. For example, in studies of brake disc rotor heating, emissivity can differ depending on whether the rotor was newly machined or covered with a transfer film from sliding against the pad material. Higher-temperature and lower-reflectivity surfaces tend to be easier to image because the thermal energy emitted is greater than that for lower temperatures and reflections from other heat sources can be discounted.

Thermography can be useful indirectly in wear diagnosis because it can sense frictional heating in bearing surfaces due to lubricant starvation, a condition that can also result in wear. Preventive maintenance literature and websites as cited in Snell [11] contain additional information and case studies. Thermal wave imaging uses an SEM fitted with an electrostatic beam blanking system, a piezoelectric transducer (PZT) under the specimen, and signal processing electronics. When a scanned, pulsed electron beam interacts with surface features, signals are transmitted by longer wavelength carriers through the specimen to the PZT. While providing information on the density of debris deposits and oxide films on sliding wear surfaces [12], thermal wave imaging has seen minimal application to wear studies.

Undoubtedly, the science and technology of imaging will continue to evolve, making new wear diagnosis tools available. The challenge of the investigator will not be a lack of tools but selecting the right tool for the problem at hand. Still, a good quality stereo macroscope with digital capture and image processing capabilities remains a good investment for those involved in wear diagnosis.

References

1. P. J. Blau, L. R. Walker, H. Xu, R. Parten, J. Qu, and T. Geer (2010) *Wear Analysis of Wind Turbine Gearbox Bearings*, ORNL Technical Report, TM-2010/59. Available from OSTI online, http://info.ornl.gov/sites/publications/files/Pub23554.pdf.
2. MicroSet™, sold by NDT Supply company, ndtsupply.com, Shawnee Mission, KS.
3. L. E. Samuels (1982) *Metallographic Polishing by Mechanical Methods*, 3rd ed., ASM International, Materials Park, OH.
4. G. F. Vander Voort (1984) *Metallography: Principles and Practices*, McGraw-Hill, New York.
5. J. F. Song and T. V. Vorburger (1992) "Surface texture," in *ASM Handbook, Vol. 18, Friction, Lubrication, and Wear Technology*, P. J. Blau, ed., ASM International, Materials Park, OH, pp. 334–345.
6. B. Bhushan (2001) "Surface roughness analysis and measurement techniques," in *Modern Tribology Handbook*, Vol. 1, B. Bhushan, ed., CRC Press, Boca Raton, FL, pp. 49–119.
7. P. J. Blau, R. L. Martin, and L. Riester (1996) *A Comparison of Several Surface Finish Measurement Methods as Applied to Ground Ceramic and Metals Surfaces*, ORNL Technical Report, ORNL/M-4924, http://www.osti.gov/scitech/servlets/purl/196529.
8. J. I. Goldstein, D. E. Newbury, D. C. Joy, C. Lyman, P. Echlin, E. Lifshin, L. Sawyer, and J. Michael (2003) *Scanning Electron Microscopy and X-Ray Microanalysis*, 3rd ed., Springer Science, New York.
9. P. J. Blau and W. A. Simpson, Jr. (1995) "Applications of scanning acoustic microscopy in analyzing wear and single-point abrasion damage," *Wear*, Vols. 181–183, pp. 405–412.
10. J. Qu, P. J. Blau, A. J. Shih, S. B. McSpadden, G. M. Pharr, and J. Jang (2004) "Scanning Acoustic Microscopy for Non-Destructive Evaluation of Subsurface Characteristics," *Proceedings of the 6th International Conference on Frontiers of Design and Manufacturing*, Paper 435, Xi'an, China, June 21–23, 2004.
11. J. Snell, "An intro to infrared thermography for mechanical applications," http://www.reliableplant.com/Read/20181/infrared-thermography-mechanical.
12. P. J. Blau and C. D. Olson (1985) "An application of thermal wave microscopy to research on the sliding wear break-in behavior of a Cu—15 wt% Zn alloy," *Proceedings of the ASME Wear of Materials Conference*, American Society for Mechanical Engineers, New York, pp. 424–431.

Chapter 5

The Tribosystem Analysis Form

The previous chapters of this book introduced the concept of a tribosystem and its boundaries, made a case for consistent terminology, provided descriptions and a coding scheme for common forms of wear, and described tools for examining surfaces and detecting wear problems. In this chapter, the tribosystem analysis (TSA) form, its structure, and its rationale are introduced. Similar in some respects to a root cause analysis, the TSA form is intended to systematically define the characteristics of specific wear and friction problems, which, in turn, can facilitate their diagnosis and suggest potential solutions.

Diagnostic tools such as the wear coding scheme and definitions from Chapter 3 will facilitate accomplishing the TSA form. That being said, a cautionary note should be restated. The multidisciplinary aspect of wear has produced component-specific terminology for naming wear types within technical communities like the bearings industry, the gears industry, the tires industry, the railroad industry, and the face seals industry. Unfortunately, such jargon evolved in a compartmentalized fashion and may not agree with the formal terminology used in the academic tribology literature or in terminology standards developed by committees of academicians and engineers. Therefore, when preparing a TSA form, it is useful not only to use standardized definitions whenever possible, but also to include the jargon of that field, if only parenthetically.

5.1 An Overview of the TSA Form

The TSA form can be customized as needed for internal company use, consulting, or failure databases. It is composed of five sections. Table 5.1 summarizes the sections. Filling in the TSA form is intended to be intuitive, but an explanation of the various sections and a few examples of entering information are also provided in this chapter.

The heading is for bookkeeping purposes. It identifies the project, one or more key technical contacts, and the date and reason for the analysis. Section 1, titled "Hardware Configuration and Materials," provides a general overview of the problem, including the physical arrangement of components. This section can include

Table 5.1 Summary of the Sections in the TSA Form

Section	Title	Entries
Heading	Project Identification and Date of Analysis	Part or component ID numbers, identification of the machinery, date, and general type of problem
1	Hardware Configuration and Materials	Description and location of the problem components or interfaces within the tribosystem, using sketches, diagrams, or photos as required
1.1	Interface Descriptions	Contact geometry, dimensions, and arrangement for one or more locations in the tribosystem; materials and treatments used on opposing surfaces; surface finishing and preparation information
2	Operating Environment	General type of relative motion and severity of contact conditions, including factors such as speed, contact pressure (load), operating temperature, lubrication method and lubricant type, third bodies (abradants, erodants, wear debris, or other), and other engineering requirements for operation
3	Problem Description	Identification of the problem details including wear type, metrics of performance, severity, and related concerns such as cost or other constraints; prior experience or history
4	Attachments and Exhibits	List of additional documents and data that may bear on the problem

one or more diagrams, photographs, and/or sketches of components to highlight the location(s) of concern. Note that Section 4 of the form can be used to reference more detailed diagrams if necessary. In practice, it is possible to have more than one wearing location within the same part of a tribosystem, or even different forms of wear can occur on the same part. If more than one problem area is in proximity to another or is experiencing similar kinds of wear, then it might make sense to indicate them as Locations A, B, and so on, within the same TSA form along with indications on the accompanying diagrams to point out their relative position. Pencil sketches, like those in Figure 4.1, may be sufficient for this purpose.

As an example of multiple wear types occurring in the same tribosystem, we consider the case of a piston ring/cylinder liner combination in an internal combustion engine. The top (compression) ring may experience sliding wear (scuffing or abrasion) against the cylinder bore on its outside diameter; the ring within its ring groove may experience microwelding due to small oscillatory motions, including fretting or "bell-mouthing" from a ring rocking action; or the liner (counterface) may experience unacceptable wear at the top dead center position. Lower down, the piston skirt may experience scuffing during cold starts and the wrist pin inside the piston body may experience wear from rapid articulation. The result of wear at such nearby locations can include loss of compression and horsepower (ring/liner), loss of efficiency, oil leakage, emissions of unburned fuel (wear volume trapped behind the rings in the ring groove), and more.

Automotive engineers have recognized the problem of location-specific wear types by using different surface engineering treatments depending on location. One manufacturer of pistons for internal combustion engines not only advertises different coating technologies used in different places on the same piston but also acknowledges that the contact conditions can change with time [1]:

> MAHLE's GRAFAL® skirt coating—reduces drag, scuffing, friction and cylinder bore wear.

And later in the same advertisement:

> Phosphate coating provides the grey appearance to the MAHLE piston. This dry lubricant coating (not to be confused with a thermal film coating) provides a lubricant film in the pin bores and ring grooves until the oiling system of the engine reaches operating pressure; particularly useful during the initial start-up or break-in of engines to protect against galling and micro-welding.

Wear at one location in a closed tribosystem can produce debris that could travel to another location and cause problems there as well. In such cases, solving one wear problem may reduce the severity or solve a related wear problem.

The interaction between wear processes on nearby triboelements from movements of debris (which the author has referred to as "tribocommunication") should be considered when defining the boundaries of the tribosystem (i.e., Block 2.8, to be described later).

Section 2 focuses on the tribological variables affecting contact conditions. That is, Blocks 2.1–2.8 provide information on the kinds of relative motion, contact pressures, surface speeds, operating temperature (or temperature range), lubricant composition and means of its supply, and the features of third bodies (external contaminants or wear debris) that might be present in the system. Block 2.9 contains important additional information. It draws attention to the need to balance materials selection by not only considering friction and wear requirements but also accounting for any other engineering functionality like fatigue life, temperature resistance, and corrosion resistance. If biocompatibility, low toxicity, or electrical insulation issues are important requirements, Block 2.9 is the place to indicate them.

Section 3 is where the problem and its possible causes are described. It answers the question posed in Chapter 2, namely: "How do I know I have a wear or friction problem?" and it supports that reply with descriptions and observations. There is space provided to indicate the dominant type(s) of wear, the metrics used to characterize the severity of the problem, any cost issues that bear on the problem, and previous experience in related cases (lessons learned on what materials or lubricants are known to work or not to work in similar tribosystems).

Block 3.2 is used to indicate the most problematic forms of wear or surface damage, and Block 3.3 indicates the metrics used to measure the severity of the problem. Examples include number of parts machined before cutting tool replacement, miles driven until a timing chain is expected to need replacement, and the critical leakage rate for a mechanical seal. Block 3.4 indicates the effect of the wear problem on machine performance. An example is the loss of compression and reduction in horsepower caused by a worn top piston ring. Block 3.5 allows the investigator to indicate how the problem is manifested. Examples include failure of surfaces to seal properly, seizure or cessation of relative motion, generation of wear debris in a food product, or wear of chain links on a heat treating furnace belt that stretches the belt and causes a food product to be out of specification (e.g., the stretching of sausages from wear of a conveyor chain as they progress through a heating tunnel).

5.2 A Further Explanation of the Entries in the TSA Form

Considering the general overview of the form and the rationale for its format, additional guidance and examples are now provided for filling in specific blocks of the TSA form, which is reproduced on the following pages.

TRIBOSYSTEM ANALYSIS FORM—Page 1

Project ID		Component or Assembly	
Manufacturer(s)		Date of TSA	
Contact person (phone/e-mail)		General Problem type	[ ] product design [ ] performance problem [ ] warranty issue [ ] other:

1 Hardware Configuration and Materials

More than one location in the tribosystem may present problems. Insert a diagram or sketch for each problem area in the machine or assembly. If the conditions are markedly different from one area to the other, preparing a separate TSA is advisable.

(Insert diagram, sketch, or schematic if attached, check here [ ] and refer to Section 4.)
Interface Location A
(Insert diagram, sketch, or schematic if attached, check here [ ] and refer to Section 4.)
Interface Location B

NOTES/CLARIFICATIONS (optional—identification of features or labels on the figures above):

_____

_____

_____

_____

[ ] continued

TRIBOSYSTEM ANALYSIS FORM—Page 2
1.1 Interface Descriptions
Interface Location A—Triboelement Descriptions

Interface (A)	Triboelement A1	Triboelement A2
1.1 General contact geometry (conformal, nonconformal, point contact, line contact, etc.)		
1.2 Triboelement shape and general dimensions (cylinder, ball, dovetail, etc.)		
1.3 Current material or surface treatment		
1.4 Surface finish on the contacting area (as-finished, as installed)		
1.5 Final finishing step (on the contact area)		

Interface Location B—Triboelement Descriptions

Interface (B)	Triboelement B1	Triboelement B2
1.1 General contact geometry (conformal, nonconformal, point contact, line contact, etc.)		
1.2 Triboelement shape and general dimensions (cylinder, ball, dovetail, etc.)		
1.3 Current material or surface treatment		
1.4 Surface finish on the contacting area (as-finished, as installed)		
1.5 Final finishing step (on the contact area)		

TRIBOSYSTEM ANALYSIS FORM—Page 3
2 Operating Environment

2.1 Type of relative motion (e.g., rolling, reversed sliding, unidirectional sliding, erosion, fretting, etc.)	A) B) (See the Note)
2.2 Speed of relative motion or range of speeds during the duty cycle	A) B)
2.3 Contact load, contact pressure, or pressure × velocity (PV) product on the contacting surface	A) B)
2.4 Temperature(s) of operation, peak/average	A) B)
2.5 Lubricants of interest (product designation, viscosity, and manufacturer or source)	A) B)
2.6 Lubrication supply method (drip, full flooded, vapor, if any)	A) B)
2.7 Lubrication regime, film thickness ratio (Λ) or region of the Stribeck curve in which normal operation takes place	A) B)
2.8 Characterization of third bodies (e.g., wear debris, grit, abrasive contaminants, if any) that appear on worn parts or are collected in the lubricant	Particle collection method/observations:
2.9 Other properties or operating requirements that are relevant to the problem (corrosion, fatigue strength, vibration, etc.)	

Note: "A" and "B" designations refer to the interface locations shown in Section 1.1.

TRIBOSYSTEM ANALYSIS FORM—Page 4
3 Problem Description

Primary operational concern(s)	[ ] friction reduction or control [ ] seizure avoidance [ ] wear reduction [ ] surface damage avoidance [ ] other: _____
3.1 Which triboelement(s) is/are experiencing problems?	A) B)
3.2 Dominant surface altering process(es) or wear modes for each triboelement (See the Note)	A) B)
3.3 Metric(s) for wear or friction that are used in the application, if any (e.g., lifetime, contamination level, clearance, visual criteria, etc.)	
3.4 Problem's impact on component performance	
3.5 Cost issues or other constraints on material or lubricant selection	
3.6 Prior experience with unsuitable materials or lubricants	

Note: ASTM G40 wear and erosion terminology preferred; enter wear type codes from Chapter 3.

TRIBOSYSTEM ANALYSIS FORM—Page 5
4 Attachments and Exhibits

Document ID	Description	Format/Filename

[ ] continued

5.2.1 A Detailed Description of the Components, Geometry, and Materials

The remainder of this chapter provides guidance and additional information on filling the various blocks in the TSA form. As noted later, it is possible that not all aspects of a friction or wear problem may be known. Still attempting to fill in as much as possible not only helps to define the problem but also indicates what is not known about it.

Section 1.1. The blocks in the first subsection on page 2 of the TSA form describe the specifics of the contact, materials, and surface finishes. Use of the term *triboelement* is made in this section as a means to identify which components are being referred to. A triboelement is defined in ASTM terminology standard G40-13(b) as follows [2]:

> *triboelement*, n.—one of two or more solid bodies that comprise a sliding, rolling, or abrasive contact, or a body subjected to impingement or cavitation. (Each triboelement contains one or more tribosurfaces.)

Therefore, a drive shaft running through a plain bearing, a piston ring, an elbow in a pipe, and one side of a mechanical face seal are triboelements. A disc brake is an example where there would be three triboelements in the tribosystem. Triboelement A1 would be the brake rotor with its outer and inner sliding surfaces, Triboelement A2 would be the inboard friction pad, and Triboelement A3 would be the outboard friction pad.

Blocks 1.1 to 1.4. Blocks 1.1 and 1.2 describe the macrogeometry, which in most cases is comprised of simple shapes. Examples include a cylinder (as in a piston wrist pin), the crowned outside diameter of a circular piston ring, or a sphere (a bearing ball). Conformal contacts are intended to refer to nominally conformal conditions and would include a cylinder inside a round journal with nearly the same diameter or a flat-faced seal rubbing on a flat counterface. Nonconformal contacts include a crowned piston ring against a round cylinder bore, a pointed set screw tip against a shaft, and a cylindrical nuclear fuel rod rubbing against a dimpled retainer. The current material (Block 1.3) for each triboelement is self-explanatory but should contain a brief description of its composition, heat treatment, source, and any surface treatments or coatings (thickness). If the material has yet to be selected for an application in the design stage, simply state: "To be determined." Surface finish (Block 1.4) may be provided using standard machining or surface roughness measurements, such as Ra (arithmetic average), Rq (root-mean-square), Rz (10-point height), and Rt (maximum peak to valley height within a sampling internal). Further information about reporting surface finishes and their designation may be found in standards documents and handbooks (e.g., Dagnall [3]). It is preferable to report the finishes of the two mating triboelements using the same set of units. In a TSA, finishes should be checked to see if the finish on the actual problem part has the same initial finish as that specified in the design. The design of surface finishes to affect functionality is a subject beyond the scope of this book, but the reader should be aware that in lubricated systems, surface finish can be key to maintaining component separation during operation. Again, if the operating surface finish has not yet been determined, as in a conceptual design, then it may be shown as: "To be determined."

Block 1.5. The final finishing step is helpful information because it can describe the lay (the directionality of the surface finishing marks) as well as the abusiveness of the process used to finish the tribosurface. Lay is important in some lubricated tribosystems in which oil retention and its movements through the interface are important. The residual machining damage may consist of near-surface residual stresses or work-hardened or textured layers from the machining process. The fact that two surfaces may have identical surface roughness parameters does not guarantee that they also possess the same degree of residual subsurface damage after machining and finishing. The second table (for Location B) may be included if there is a nearby or related tribosurface. For example, if the upper piston ring on bore is Location A, then Location B might be the region where the piston skirt rubs against the cylinder bore. Both of these are reversed sliding locations and they share a common partner (the cylinder bore), but the geometry and temperature of operation may be quite different. If, on the other hand, the piston pin on its connecting rod's small-end bearing were also of interest, then preparing a separate TSA form for those triboelements would be a better choice since their basic geometry and conditions of operation are significantly different.

5.2.2 A Description of the Operating Conditions

Section 2 of the TSA form lists the operating conditions for the tribosystem. These include the type of relative motion, relative speed, the applied forces, temperatures, and other functional characteristics.

Block 2.1. The type of relative motion of the triboelements refers to whether, for example, their surfaces move unidirectionally, with reversed motion (reciprocating), with normal components as well as sliding components (as in impact wear), or by some kind of compound motion, like the articulation of a ball-in-socket joint in a normal human being during her daily activities. As indicated in the title of Block 2.1, the type of relative motion can be further described with terms like fretting, rolling with slip, or other specific descriptors. Furthermore, in the case of fretting, four types of motions can be indicated (see Figure 3.20). If the motion is intermittent, that can be indicated as well.

Block 2.2. This describes the speed(s) of relative motion and is applicable to sliding (with or without abrasion) or rolling contact between mating surfaces. If rolling contact is involved, information on the slide/roll ratio or percentage slip can be added here. In some machine configurations, the operating speed is given in terms of revolutions per minute (rpm), but in this case, the radial distance of the tribological interface from the center of rotation should also be given to enable a relative interfacial velocity to be calculated. If erosive wear is involved, this is the place to insert the flow velocity or flux of the erodant.

Some equipment runs up to speed and remains at that speed. In other cases, like brakes and turbochargers, the speed varies widely during operation. If the system is oil lubricated, then the effectiveness of the entrained lubricating oil in creating a separating film between surfaces is dependent on entrainment speed among other factors (see also Section 3.4). Therefore, changes in operating speed can affect the degree of surface contact from none to full solid–solid contact. Consider a piston ring in an internal combustion engine that varies in speed from zero at the turn-around points to a maximum at the middle of the stroke. Wear tends to be greatest at the top dead center and at the bottom dead center.

Many of the traditional types of wear tests, like those contained in the ASTM literature (e.g., pin-on-disk, block-on-rotating ring, dry sand/rubber wheel abrasion) are operated under constant speed conditions. Even step-loading tests, in which the applied force is raised gradually until the load-carrying capacity or galling resistance of a surface is exceeded, tend to be run at constant speed. The fact is that many practical tribosystems do not operate at constant speed and, further, the cumulative rate of wear can be more influenced by the number of start-up/shut-down cycles (non-steady-state conditions) than the time spent under steady-state operation. Therefore, testing methods should reflect those transient conditions, and Block 2.2 in the TSA is the place to describe such conditions clearly in the definition of the tribosystem.

Block 2.3. Practically speaking, there is confusion in the literature over four terms that are often used to describe the magnitude of the mechanical conditions that are applied to hold wear and frictional surfaces together: (*a*) load, (*b*) normal force, (*c*) contact pressure, and (*d*) contact stress. Terminology can cause some confusion when describing a tribosystem because certain key information may be missing from such descriptions. For example, a force is applied to an apparent area of contact to produce an apparent (a.k.a., nominal) contact pressure. The apparent contact pressure is not, in general, the localized pressure on the microscale or nanoscale asperities on that surface. The only case in which that would happen is if there was complete solid–solid contact, with, in effect, one asperity stretching all the way across the entire surface. Likewise, the true area of contact, the sum of all the points in the interface where physical contact exists between opposing bodies, is well known from the classical tribology literature (Bowden and Tabor) to be, in general, smaller, sometimes very much smaller, than the apparent area of contact. Hence, the true contact pressure, which is sometimes equated to the indentation hardness when surfaces are at rest, may be considerably higher than the apparent contact pressure that is computed from the applied external load and the apparent bearing area.

When completing a TSA Block 2.3, consider the following differences and uses of the terms load, force, pressure, and stress:

1. *Load (W)*. The term *load* is somewhat ambiguous because it can refer to a weight bearing downward on a surface, as in "I applied a load of 20 kg to the sliding pin." The downward force then, if the sliding surfaces are relatively moving perpendicular to the gravitational pull of the Earth at that location, is equal to 9.807 m/s^2 × 20 kg, or 19.61 N. However, some authors report load in units of Newton (N). Therefore, there is some ambiguity, and the context of the statement must be ascertained. A load is usually applied downward and toward the earth. Such ambiguities can be resolved by instead using the term *normal force*, which acts normal to a bearing surface irrespective of how the surface is oriented with respect to the gravitational pull of the Earth.

2. *Normal force (F_n, sometimes designated as P)*. This term is preferred over *load* because it is less ambiguous and refers to the magnitude of the force applied perpendicular to the contact surface. If the surface is curved, then the normal component of the applied force will be different at different locations on the curve. The friction force or tangential force at the location of interest can also be listed if that information is available. If the magnitude of these forces is calculated from the machine design, measured in some way, or estimated, then a note indicating the manner of force determination can be added to the value in Block 2.3. It is also possible that a range of forces is experienced by the component, and this can be noted here as well.

3. *Nominal contact pressure (p_{nom})*. The normal force divided by the apparent (nominal) area of contact of the mating components is called the nominal contact pressure. Pressure is expressed in the same units as stress but is not necessarily equivalent to stress (see item 4). It is relatively easy to estimate nominal pressure from the design and macrodimensions of the contact surface. In counterformal surfaces, bear in mind that during operation, wear may cause one surface to change its area, and therefore, p_{nom} may change with time. In addition, in some machines, P is a function of time. The force on an internal combustion engine's piston rings and the forces on a friction brake are examples of normal forces that change during operation. Stating the typical maximum, minimum, and operating average value of p_{nom} is helpful in a TSA. Design and material selection philosophy can be affected by a decision as to which value of the contact pressure should be used for tribosystem design (i.e., design for the average or design for the maximum?). In brake material testing, that industry has developed standardized, computer-controlled testing protocols that apply a series of stages of contact severity (speeds, contact pressures, and temperatures) to simulate various driving conditions and types of vehicles.

4. *Contact stress (σ_c)*. Contact stress is expressed in units of force/area, like pressure. Stress is the magnitude of the response of a solid body to a force imposed on a designated area within or upon it. Technically, therefore, one can apply a force but not directly a stress. It is up to the material to react to the force applied in a certain direction by becoming stressed on an area perpendicular

to the force application direction. Often, the term *contact stress* is used by bearing designers to refer to the maximum Hertz stress, which is computed from equations of elasticity and may be found in numerous tribology handbooks, texts, and basic contact mechanics. The Hertz stress, which should be further qualified when cited as to whether it is the maximum compressive stress or some value other than that, say, the average contact stress, is based on recoverable elastic deformation and, as such, would not apply if the forces on the contacting bodies cause them to deform plastically and irrecoverably. Perhaps, a certain location within the tribosystem is designed to operate under a given maximum Hertz stress ($\sigma_{H,max}$); that value could be inserted in Block 2.3 and designated as the design stress. But if wear occurs and the contact changes size or shape, the actual operating stress may be other than that value.

Design calculations of loads or pressures may not always be available for the specific location where wear has occurred. For example, when designing a tribotest for selecting a housing material around the gear in a gerotor-type hydraulic fluid pump, designers could very accurately compute the contact forces and pressures on the gear teeth within the housing but were flummoxed when asked for the contact pressure to the *sides* of the gear that spun against the housing. This value required some educated guesses about the pressurized fluid that leaked around the opposing sides of the gerotor gear, but it was not straightforward. Ultimately, the magnitude of the normal force used in laboratory tests of candidate housing materials was based on replicating the type of wear surface features that were observed in torn-down pumps.

Block 2.4. The temperature refers to the temperature in the space surrounding the tribocontact. The normal operating temperature is the temperature of the surroundings plus any contributions due to tribological contact (frictional heating). This may be the normal operating temperature, but it can note the range of operating temperatures as well. Some bearing components may run hotter during the wear-in period and then settle down to a steady-state temperature. Likewise, an automobile engine may experience periodic cold starts that affect the properties of the lubricants and, hence, the wear rate. It would be appropriate to note such temperature variations.

Blocks 2.5 and 2.6. These blocks identify any lubricants used and the method of their application. The trade names or compositions of any lubricants used in the tribosystem, solids or liquids, should be given in Block 2.5. Generally, if the system depends on the formation of lubricative oxides during operation, that can be mentioned here as well. Block 2.6 is the place to indicate how the lubricant(s) is supplied. For example, light-duty bearings may simply use a solid lubricant that is sprayed on during assembly with no more added. Perhaps, the lubricant is built into a composite bearing material (e.g., polytetrafluoroethylene [Teflon] containing "self-lubricating" polymer composites) or it may be infused into a porous bronze bearing. In contrast to such approaches, a lubrication system can be a complex loop system involving metered nozzles, feeding lines, multistage filters, coolers, and built-in diagnostic bypass loops. The design of a lubrication system can be influenced by such factors as the accessibility of the tribocomponents to be lubricated (say, bearings used in spacecraft or satellites), the volume of lubricant required, the method used to feed the lubricant into the tribocontact, the need for filtration, the need for cooling, the sharing of a lubrication system between more than one tribocomponent in the machine, and many others. It is possible that harmful third bodies are entering the wearing contact as a result of wear occurring elsewhere in a neighboring component that happens to share the same lubrication system. The process by which the wear of one component is accelerated by debris from another has been referred to as *tribocommunication* [4].

Block 2.7. One important source of wear transition is a change in the lubrication condition. The general subject of lubricated wear is a vast one, and its discussion falls well beyond the scope of this book; however, it is worthwhile introducing the concept of lubrication regimes because failure of a lubricant to perform as planned can result in transitions from a low rate of wear (say, light 3-body abrasion from entrained particles or debris) with negligible surface contact to a much worse situation (e.g., scuffing or adhesive wear).

Basically, there are three lubrication regimes discussed in the literature. The Martens-Stribeck curve forms the foundation for much of the interpretation of lubricated sliding and wear behavior, and for a more detailed discussion of lubrication regimes, the reader is referred to Jackson [5]. In summary, the three lubrication regimes are (i) boundary lubrication, (ii) hydrodynamic lubrication (HDL), and (iii) mixed film lubrication that falls between the first two regimes. These are based on the existence of a fluid film that either fully separates relatively moving surfaces or allows them to touch. If the fluid film allows opposing surfaces to touch all or most of the time, then we say there is boundary lubrication. A thin layer of fluid or chemical reaction products may coat the asperities on the wear surface during boundary lubrication and provide some level of friction reduction and wear protection. Even worse, if there is no lubricant present, we can call this a "starved" condition and expect to find the kind of wear and friction levels characteristic of bare, dry, sliding surfaces. In contrast, if the fluid flow into the contact creates sufficient pressure to force mating surfaces apart so that wear does not occur at all, then we call this hydrodynamic or "full-film" lubrication. Sliding bearings are commonly designed to produce full-film lubrication.

The so-called Stribeck curve (originally developed by A. Martens in the late 1800s and popularized by R. Stribeck several decades later), represents the relationship between the relative surface velocity (v), the applied load (P), the viscosity of the lubricating fluid (h), and the resulting friction coefficient (μ) for a journal bearing (turning shaft in a sleeve). The concept has been extended to other contact geometries as well. Basically, the concept involves the proportionality:

$$\mu \propto \frac{v\eta}{P} \tag{5.1}$$

It also embodies a concept known as the film thickness ratio or lambda (Λ). The lambda ratio is the ratio of the oil film thickness in the contact to the root-mean-square roughness of the two mating surfaces. Therefore, if Λ is large, there is more separation and less wear.

Figure 5.1 displays the general form of a Stribeck curve showing values for Λ under different regimes. Note the curve's asymmetry and the rapid rise in friction to the left corresponding to a rise in normal load or a decrease in either speed or viscosity. Stribeck-based lubrication condition analysis can be powerful, leading to insights into the stability or transitions of sliding conditions. Note that the optimal combination of load, velocity, and viscosity for minimal friction can be found in this representation. Note also that changes in surface finish can shift the position of this curve. Surface textures, like patterns of shallow dimples or channels of various shapes, can also shift the position of the curve.

If the operating point of the tribosystem on the Stribeck curve shifts in some way during normal component operation, then friction and wear can change

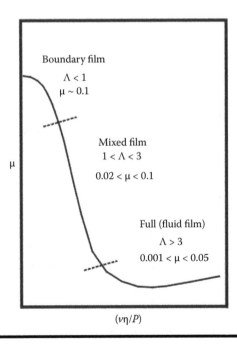

Figure 5.1 **A typical Stribeck curve shape showing the possible regimes of a lubricated tribosystem's operation. Sliding of very clean surfaces can produce friction coefficients well over 1.0, so the left axis of the figure implicitly assumes that some level of lubricant is still present.**

accordingly. Such a shift can occur gradually, as when a lubricant ages and its viscosity changes over time. But the operating point can shift periodically up or down the curve if the surface speed, like that of a piston ring against a cylinder liner, which varies as the crankshaft turns. Portions of the piston ring stroke can be in boundary regime (top dead center or bottom dead center), while other portions can be in mixed-film or HDL (midstroke). Therefore, wear transitions can occur rapidly and periodically in some tribosystems, with the net effect of this being to promote wear at some locations but not at others. For example, it is common to observe the most wear on the cylinder bore in an internal combustion engine at the upper end of the ring stroke (top dead center), where the high heat and collapse of lubricating films lead to more damage. Finally, the cold start-up of sliding machinery or other temperature-induced changes in lubricant properties can produce different lubricating regimes and, hence, a greater chance for wear. Similarly, the use of different surface treatments can rank in opposite order of wear resistance depending on the type of wear (see Section 6.2).

Block 2.8. The presence of third bodies generated within or external to a tribosystem can surely affect wear. Chapter 2 described methods for collecting and characterizing third bodies. If third bodies are present in the system, this is the place to describe what they are, how they are collected for analysis, and what their composition, size, and appearance suggest about their origins. As in the case of fretting, adhesive wear, and abrasive wear, the nature of the third bodies are indicative of the type of wear taking place. Techniques like ferrographic analysis, filtration, and scanning electron microscope (SEM) examination can shed light on the sources of third bodies.

Block 2.9. Most bearing materials and lubricants are selected based on a balance of required properties and characteristics. For materials, these include such diverse considerations like minimum tensile strength, elastic modulus, hardness, corrosion resistance, cost, availability, heat resistance, thermal conductivity, coefficient of thermal expansion, fracture toughness, fatigue strength, damping capacity, appearance, familiarity, and machinability. For lubricants, these include viscosity characteristics, oxidational stability, pourability, detergency, antifriction and antiwear characteristics, and more. Sometimes, a compromise in one property is needed to meet the targets established for another. For example, in order for a certain steel to have a minimum fracture toughness level, its hardness level may need to be compromised by a few points. Or in order for a certain corrosion level to be met, more expensive alloys may be required and the cost compromised.

5.2.3 A Problem Description

The previous sections of the TSA form describe the normal operation of the tribosystem and the environment of one or more of its triboelements. Block 3 of the form is where the actual problem is described. The top block (not numbered) begins with a statement of what category the problem may fall into, such as friction reduction, wear reduction, or other concerns. Examples of "other" concerns are specific situations such as "heat checking," microwelding, micropitting, and fretting fatigue initiation. If these are specifically called out here, they should be consistent with the entrees in Block 3.2.

Block 3.1. This block identifies which triboelement(s) are experiencing the problem. There is space to indicate components for Locations A and B. Obviously, this block can be modified as required to add more or fewer wear locations and should be consistent with the information in the summary section, 1.1.

Block 3.2. This block is intended to identify the primary and secondary wear types of concern. The wear category codes listed in Figure 3.3 provide shorthand for indicating these, but additional details may be added, including any terminology common to the industry. Any detailed documentation such as images, field notes, or relevant details can also be listed and referenced in Section 4. A footnote in Block 1.1 may be placed here to refer to those sources.

In addition to documenting the form of wear based on its observed characteristics, indications of the severity of the wear can also be inserted here. For example, when indicating the severity of rolling contact with slip, the hierarchy shown in Chapter 3, Figure 3.37, could be used.

Block 3.3. The quantities used to measure wear or frictional behavior of concern in the tribosystem or subtribosystem of interest are indicated here. Unlike the kinds of fundamental quantities that are measured in research laboratory wear tests, these may be either direct or indirect measurements, such as those discussed in Chapter 2. Examples include the following:

- Leakage rate of a seal over time
- Extrusion back pressure or output of a plastic extrusion machine
- Thickness of a disc brake pad
- Loss of compression in an internal combustion engine
- Clearance in a plain bearing
- Vibration level in a bearing
- Electric current draw by a motorized valve gate
- Temperature rise in a bearing
- Scoring marks on a rolled metal sheet (product quality reduction)
- Torque on a lead screw, a clutch, or a brake
- Friction fall-off during braking of a vehicle ("fade")

Methods to indicate the quantitative or qualitative degree of wear (such as the exposure of built-in wear indicators in tires or brake pads) can also be listed in Block 3.3 and used as a metric for acceptable or unacceptable performance.

Problems in field/lab wear testing correlations can occur if the metric used in the field (e.g., thickness of a brake pad) is not the same as that used in the laboratory to screen materials for that application (e.g., weight loss of the sliding "pin" specimen). Such problems can be avoided if full-sized machinery or a production-sized simulator is used to conduct tribotests for material or lubricant screening, but that is not always cost-effective or practical. As often occurs, laboratory friction and wear tests use simpler configurations than the field application and may produce metrics other than those used in the field. Again, weight loss or wear track depth may be convenient to measure in the laboratory, but thickness reduction may be a more convenient measurement in the full-sized field component. Herein lies one of the major challenges in applied tribotesting, namely:

> How closely and in what respects must a tribotesting apparatus ("tribosystem 1") resemble the tribosystem of concern ("tribosystem 2") in order to produce data that are useful for material or lubricant screening, ranking, and selection?

A block-by-block comparison of the TSA for the testing machine with that for the field component can determine how closely the tribotester simulates the field component, but even so, aspects such as the energy dissipation from the contact interface and the motion of third bodies may be difficult to simulate in the laboratory without considerable effort (if at all). One way to address the dissipated energy question is to examine the wear surfaces and temperatures generated. Another way is to develop a design metric involving some of the variables that contribute to energy sources in the interface. Examples include the PV limit that is described in the following paragraphs.

In some design applications, notably, plastic bearings, bushings, face seals, and sleeves, the maximum value encountered of the product of the nominal contact pressure (here labeled p_{nom}) and sliding velocity is used as a design and material selection criterion. This so-called PV limit, which is generally determined by running a matrix of experiments at different sliding pressures and speeds, can be used to serve as a limit for the qualification of candidate materials (plastics and polymer compositions). That being said, an important cautionary consideration is that the test selected to measure the PV limits of candidate materials should simulate the conditions of the application as closely as practicable.

The product of the nominal contact pressure (p_{nom}), sliding velocity (v), and the kinetic friction coefficient (μ) gives the applied frictional work (e.g., N-m or J) during sliding U_F:

$$U_F = \mu p_{nom} v = \left(\frac{F_f}{F_n}\right)\left(\frac{F_n}{A}\right)v = \left(F_f/A\right)v \qquad (5.2)$$

Note that U_F is the frictional work per unit contact area per unit time. The energy arising from frictional work may be dissipated differently in different tribosystems. Part of it may be conducted to the surroundings, generate vibrations, generate wear particles and deformation, or be stored in the material (cold work in metals or phase changes in these and other materials). If the application of interest does not partition the available energy in the same way as the testing apparatus used to determine the PV limit, then the use of PV criteria in Block 3.3 may require modification to enable more relevant material screening.

There are, of course, a number of possible metrics besides the PV limit that could be used to compare friction and wear performance depending on the configuration of the tribosystem. Two examples are (i) the "work rate" used in designing nuclear heat exchanger and core components to avoid fretting wear [6] and (ii) the "frictional heating parameter (Φ)" introduced by Blau et al. in 2007 [7] to characterize the ability of experimental friction brake disc materials to dissipate heat under similar sliding conditions. The latter (measured in degrees Celsius per Joule) was defined as the frictional work required to raise the measured temperature on a sliding surface of a subscale brake disc (measured by infrared [IR] thermography) by a given amount based on the average friction coefficient and sliding distance:

$$\Phi = \frac{\Delta T}{\mu P v t} \tag{5.3}$$

where ΔT is the measured temperature rise due to sliding, μ is the average friction coefficient over time interval t (seconds), P is the applied force (Newtons), and v is the sliding velocity in meters per second. Comparing under the same sliding conditions of constant drag speed and force, for example, a commercial semimetallic brake pad material against cast iron gave $\Phi \sim 1.2 \times 10^{-3}$ (°C/J). That for the same pad against less thermally conductive Ti-6Al-4V was $\Phi \sim 8.3 \times 10^{-3}$ (°C/J). Adding a conductive phase to create a composite of Ti-6Al-4V (30 vol% TiB_2) lowered Φ to $\sim 4.4 \times 10^{-3}$ (°C/J). Here, a metric was created to enable screening of candidate brake lining and disc material combinations.

Block 3.4. This block is used to describe the effects of the wear (or friction) problem on the performance of the tribosystem. The description may not need to be lengthy. Examples include loss of efficiency, premature wear-out, inability to remove fasteners, loss of compression or sealing, and customer dissatisfaction with appearance. Block 2.9 can be referred to if the other engineering requirements are degraded by the wear problem. For example, wear of a metal mixing paddle could introduce debris particles into a food product or the wrong lubricant choice for can necker dies could change the taste of canned beer.

Block 3.5. This block allows the user to enter the additional considerations, other than tribological ones, that bear on the problem at hand. Examples include the location of the machinery in hot and humid places, the difficulty of avoiding contamination from the surroundings, the vibration of adjacent equipment, the need to use certain vendors or limited choices of options in lubricants, materials, and surface treatments. These are constraints on materials, surface treatments, or lubricants solutions. An interesting case is the acquisition of clean laboratory space for executive staff offices that caused wear testing instruments to be moved to a dirtier location in the plant and a compromise in the quality of wear data.

Block 3.6. Lessons learned and problem history. The current problem may be part of a history of wear problems of this tribosystem. Perhaps, an earlier problem was addressed by a "quick-fix" and the nonoptimum results of that attempt are coming back to create more problems. It is also possible that a similar problem in another tribosystem was successfully treated. If a consulting tribologist or experienced engineer is brought in to help with the problem, it is helpful to probe his or her experience in dealing with similar types of problems and to indicate what has been tried before (both successfully or unsuccessfully). Knowing what does not work is nearly as important as knowing what does work.

5.3 Missing Information

When developing a TSA form, it is likely and even probable that some of the information that one would like to include about the materials, the mechanical configuration, and/or the operating environment of the tribosystem is not available. Perhaps, the TSA form is being used for an untested design prototype, or the information from the field is incomplete. Perhaps, the surfaces of the worn components returned from the field were not well preserved for examination. Either they were allowed to corrode after removal from the machine or they were cleaned and important clues (like adherent debris) were lost forever. Maybe worn field components were discarded and no longer available for examination. Conceivably, the component of interest operates under a variety of non-steady-state conditions (e.g., the brakes of an urban delivery truck with different drivers and routes) and the notion of an average or maximum load or speed is an oversimplification.

Missing information, especially that having to do with operating conditions, has implications when selecting test methods to screen materials or lubricants. The concept of a nominal tribological scenario (NTS) or maximum severity tribological scenario (XTS) can be considered. These terms are defined as follows:

- *Nominal tribological scenario* (NTS)—In a system that involves relative motion and the contact of surfaces, those mechanical, thermal, and environmental conditions that exist during the average or most probable operating conditions of that tribosystem.
- *Maximum tribological scenario* (XTS)—In a system that involves relative motion and the contact of surfaces, those mechanical, thermal, and environmental conditions that exist during the most severe or intermittent operating conditions likely to be encountered by that tribosystem.

The NTS does not need to consist of a single set of constant load, speed, temperature, and exposure time. In fact, some industry methods for testing friction and wear involve a sequence of stages (steps) designed to simulate the typical variations in service conditions. One example is the use of standardized, computerized test procedures developed by organizations in the friction brakes industry (e.g., Society of Automotive Engineers [SAE], International Standards Organization [ISO], and the Technology and Maintenance Council [TMC] of the American Trucking Associations). These are used in dynamometer testing of brake linings (a.k.a., "friction materials"). As many as 8 to 10 stages involving various speeds, temperatures, application times, and contact pressures are applied in order to evaluate characteristics like nominal friction coefficients, effects of speed and deceleration on brake fade and recovery, and of course, wear. Such procedures may be aimed at replicating conditions of the type of service the vehicle may experience. Multistage programmed tests for friction materials can take over a week to complete. These include scenarios to simulate use during long-haul highway service, city driving, delivery trucks, fire vehicles, and municipal buses. While such tests provide a great deal of information, they are also expensive to run, and that fact reduces the ability to establish the repeatability of the test results by running each experiment multiple times.

The XTS presumes that materials must be able to withstand the most severe anticipated operating conditions in the tribosystem and establishes a target pass–fail condition. An example of that is the PV limit often used for selecting polymers for bearings (see the description of Block 3.3). Another example is the maximum operating conditions of the vane pivots in a variable turbocharger. Depending on engine design, operating conditions can range from cold starts to episodes of running at temperatures in excess of 600°C. Under such severe operating temperatures, no liquid lubricant may surface and the components may have to depend on lubricative oxide scales that form in situ to protect the components and reduce running friction.

XTS need not involve high loads or high operating temperatures. In fact, some of the most severe conditions of an XTS can involve low loads and low temperatures. Two examples of this phenomenon are "off-brake wear" of disc brakes and cold starts in internal combustion engines.

Brake pads are designed to retract from sliding contact when the brake is off, but due to mechanical problems, the pad may fail to separate from the rotor when the brake is off, and continuous sliding can occur with a small but finite load still applied. This is called "off-brake wear" and can be studied by running at low loads for an extended period of time. Cold starts are commonly blamed for wear problems in internal combustion engines like piston skirt wear. The engine has not yet warmed up to its optimal operating conditions, and key contact surfaces in the engine may not yet be fully lubricated. Automotive industry studies have shown that frequent short trips tend to be more damaging to engines than running for longer distances without such stops and starts. Therefore, an XTS approach to testing could involve not necessarily long test times but rather impose many cold starts or stops in which the lubricant has not achieved a full film condition needed to protect the components.

The operating conditions in Section 3 of the TSA form are a key to diagnosis because including such information as transients as well as normal operating parameters can be invaluable. In practical machinery, wear need not progress at a constant rate, so much of the lifetime of certain components can be spent under very low wear conditions. Avoiding high wear during running-in, transients, or premature wear-out may hold the answer to understanding and solving such problems.

5.4 Tailoring the TSA to Practical Problems

While similarities can and do exist between various types of practical tribological problems, each new problem may comprise a set of slightly different constraints and operating considerations. The TSA form helps structure and define wear (and friction) problems, but it may need to be modified to suit whatever aspects of the current case requires special attention or emphasis. The following case studies exemplify the need to tailor individual TSAs and use them to select test conditions.

5.4.1 Piston Ring Groove Wear

A company decided to lightweight the pistons in its internal combustion engines by installing wear-resistant inserts at the top of the piston. The upper compression piston ring is seated in a groove that must have minimal crevice volume behind the ring or unburned fuel may become trapped there and eventually expelled with the exhaust. Wear of the lightweight composite piston ring groove against the ring, a harder material, can occur due to the compound motions of the ring groove. The ring may rock to cause bell-mouthing of the groove. It may move in and out of the groove in a fretting motion. It could impact the top and bottom of the groove walls. It can rotate (slide) in the groove in one direction, then reverse and slip in the opposite direction. Microwelding, a form of adhesive wear, was especially to be avoided.

TSA Considerations: Emphasis is on elevated temperatures, likely lubricant starvation in the contact, compound interfacial motions, and modes of groove deformation and wear: ring rocking, fretting, slip, and impact. The objective was to design a laboratory screening test to simulate this type of wear and screen material combinations; there was a focus on simulating the type of wear observed in pistons. An ultrasonic vibrator (voice coil) was used to cause piston ring segments to oscillate within a cartridge-heated section cut from the upper ring groove of a production piston. After testing, the interior of the ring grooves were examined to confirm the presence or absence of microwelding and adhesive wear damage, which simulated the kinds of features seen in grooves from pistons from fired engines. Laboratory coupon tests of piston grooves with production rings led to a down selection of materials for subsequent engine testing.

5.4.2 Vanes for Diesel Engine Turbochargers

One type of turbocharger design involves moveable vanes to control gas flow as the engine's power demands vary. Depending on the design, temperatures higher than 550°C can be present in the environment where the vane ends pivot in the ring in which they are seated. Liquid lubricants are impractical and solid lubricants applied during manufacturing or service could be worn quickly away. Use of oxides formed on the vane and ring materials to lubricate the pivots was a major consideration in material selection.

TSA Considerations: The duty cycle of the components, including temperature excursions, had to be captured in the TSA form. Potentially lubricative oxides that form at different operating temperatures vary in chemistry, crystal structure, and tribological properties. It was important to capture the type of motion (pivoting), the operating temperature profiles, and the composition of the gases in the turbo-charger loop when defining the tribosystem.

5.4.3 Diesel Engine Valve/Valve Seat Wear

Exhaust valves in diesel engines see somewhat different operating conditions than intake valves do. Both poppet valve stems slide up and down rapidly in their guide tubes, which are lubricated by whatever oil manages to find its way into the guide bore. The mating valve seats, which are commonly inserts of wear-resistant alloys, experience a combination of impact and slip as the valve closes. If small channels develop around the periphery of the valve bevel, then a phenomenon called "torching" can occur as erosion allows gases to escape locally through these grooves. The exhaust valves tend to run hotter, as hot as 850°C, accelerating the role of tribo-oxidation, heat checking, and related thermal damage.

TSA Considerations: Impact plus slip occurs as the valve engages the seat. The contact pressure can be estimated based on compression pressure in the fired engine and the contact area around the valve bevel. Some valves rotate around the stem axis and others do not. Since there is a potential for impact, slip, and later erosion (torching), the relative extent of each wear damage mode for the given engine design needs to be identified by examination of valves from the field.

5.4.4 Lightweight Rotors for Truck Brakes

Brake discs engage their pads of "friction material" during snubs (decreases in vehicle speed) and full stops. The energy generated by friction can be enormous. It was estimated that one stop of an 80,000 lb, class 8 truck from 60 mph to 0 mph could generate enough friction heat in the brakes to heat the average-sized Michigan home for one winter [8]. During the last decade of the 1900s and the early decade of 2000, there was an interest in light-weighting vehicle brakes and providing them with added resistance to corrosion from aggressive $MgCl_2$ and NaCl melting compounds used on icy roads. Materials such as aluminum alloy–SiC composites found their way into some automotive brakes, and there was some interest in titanium alloy brakes for heavy trucks. The two alloy systems had quite different responses to heat. The aluminum alloy had a relatively high thermal conductivity relative to cast iron brake materials and ran cooler. However, their friction against typical brake pads was borderline too low for effective braking. On the other hand, the titanium alloys have about 1/7 the thermal conductivity of cast iron and tend to run much hotter even though they have better frictional characteristics and resist fade after repeated braking events.

TSA Considerations: Since the duty cycle affects fluctuations in temperature of the sliding interface, such things as how much time there is between subsequent braking events (highway travel versus in-town travel), the terrain (flat roads, hills, and mountain descents), weather (temperature swings and whether the surfaces are wet), and driver behavior all affect the braking environment. Therefore, the range of operating speeds, temperatures, and interfacial contact pressures is not easy to define. Some auto companies have instrumented vehicles in various geographic locations in an attempt to produce typical usage spectra. Vehicle uses must therefore be taken into account. That is why industry, the government, and trade organizations have developed multistage friction and wear dynamometer testing protocols.

5.4.5 Gerotor Cases for Fluid Pumps

A gerotor pump consists of an eccentrically mounted spinning gear (gerotor) that traps incoming liquid (like transmission or hydraulic fluid) and pushes it out at higher pressure. The sides of the gear spin against a housing. Alternate materials with lighter weight or lower cost but with increased durability were desired for the housings. Designers had extensive computer codes available to model the wear and stresses of the gerotor teeth but could provide very little information on the pressure holding the spinning gerotor against its case.

TSA Considerations: Lack of design information on the contact forces on the sides of the gerotors required more attention to the modes of surface damage. Abrasive wear (2BAb, 3BAb) with occasional scuffing was observed. The load used in simulative screening tests was determined by their ability to create the forms of wear observed on field components.

5.4.6 Fuel Injector Plungers for Low-Sulfur Fuels

Reducing sulfur emissions was important, as energy efficiency of cars and trucks must go hand in hand with emission reduction. Low-sulfur fuels were believed to reduce fuel lubricity and increase wear in engine components like plunger-in-bore types of fuel injectors. The diametral clearances in commercial fuel injector plunger/bore assemblies can be as small as a micrometer or two, so the plunger and bore had to have similar coefficients of thermal expansion and reduced propensity for scuffing. Attempts were made, with some success, to substitute ceramics for metal or hard-coated metals in fuel injector plungers.

TSA Considerations: The primary form of surface damage was scuffing that led to 3BAb wear or seizure. Contact conditions were difficult to compute because while the velocity was well known, the effects of tiny axial alignment errors on localized contact between plunger and bore were difficult to model. One approach to testing is to study the progression of wear damage (scuffing) in reciprocating tests (crossed-cylinders tests), using high-speed data capture friction force information captured for the initial back-and-forth stroke and compare it to periodic samples of an up and back stroke after running for various times. Differences in friction imply changes in the state of surface damage, and the time to initiate and propagate scuffing damage can be monitored.

5.4.7 Wind Turbine Gearbox Bearings

At this writing, tower-based wind turbines have been scaled up from kilowatt-sized units to some capable of providing 5 MW or more of electrical power. The cowling at the top of the tower that houses the gearbox and electronics can be as large as a small bus. Of course, rising power outputs place more demands on the moving parts. Common designs involve a large rotor (40 m or longer blades), a high step-up ratio (1:30 or 1:40) gearbox, an electricity generator with a control package, a system to keep the rotor facing into the wind, brakes, and a supporting structure. The reliability of such lubricated gearboxes is one of the most important enabling challenges. Turbines are expected to produce electrical power reliably for 10–20 years or more. The Internet reveals a number of tribology problems (both friction and wear) and reliability challenges associated with the bearings in wind turbine gearboxes.

TSA Considerations: A wind turbine gearbox is a complex, multicomponent tribosystem with numerous triboelements (bearings, gears, seals), and it would therefore be impractical to try to capture its details on a single TSA form. As with any complex mechanical system, the design and operating envelope determines the loads, motions, speeds, temperatures, and lubrication requirements of wind turbine gearbox bearings. In order to define such a tribosystem, the specifics of each wear problem should be captured in its own TSA form. There is a need to characterize a range of operating temperatures, contact pressures, and as-installed or run-in surface finishes (a key to micropitting control). Accelerations and decelerations during fluctuations in the wind may be problematic because when gear and bearing speeds change, the ability of the lubricating films to separate surfaces is compromised. It is proposed that transient testing and a study of transients in load, speed, and lubricant temperature, rather than the steady-state performance, may have the highest correlation with reliability of such tribosystems.

5.4.8 Human Teeth

Wear of tooth enamel and dental restoratives comprises a complex tribosystem in a tribochemical environment (saliva). Chewing motion and forces vary with the individual, and there have been a variety of standard test methods developed to simulate dental material wear.

TSA Considerations: Part of the problem in conducting a tribosystem analysis and in the selection of appropriate test methods for dental restoratives is the wide range of loads. For example, nominal bite forces when chewing food range between 10 and 120 N, but maximum forces can be as high as 190 to 290 N [9]. Motions can be represented by a half-sine wave with a frequency of 0.2 to 1.5 cycles per second. About 1,200,000 cycles must be applied in testing to simulate 5 years of chewing. The types of wear codes include 2-body repeated impact (RI/2B), 3-body abrasive wear (S/3BAb), and multidirectional sliding (S/MD). An article by Lambrechts et al. [10] lists the types of dental material wear as being (1) 2-body abrasive wear, (2) 3-body abrasive wear, (3) adhesive wear, (4) fatigue wear, and (5) tribochemical wear ("dental erosion"). The fitting of upper and lower features of the teeth can produce high stress concentrations at the high points on opposing cusps. The used gold crown shown in Figure 5.2 indicates plastic deformation at the high spots and some minor abrasion at a few other locations.

As the intricacies of specific wear and friction challenges come to light, completing a complete documentation of some complex engineering tribosystems may begin to seem out of reach. However, numerous practical wear and friction problems have been successfully treated in the past by the application of logic, engineering acumen, educated reasoning, experience, proper testing, trial and error, and luck. By preparing a TSA, it is hoped that the latter two approaches can be avoided.

Figure 5.2   This gold crown was lost from a third molar after about 2 years of use. Other than a few minor scratches and some plastic deformation of the ridges surrounding the cusps, the wear was sufficiently low to enable its reinstallation. Scale units are mm.

References

1. Mahle, Inc., www.us.mahle.com/mahle_north_america/en/motorsports/proseries -pistons/ (accessed September 30, 2015).
2. ASTM G40-13 (2013) "Standard terminology for wear and erosion," in *ASTM Annual Book of Standards* Vol. 03.02, ASTM International, W. Conshohocken, PA, http://www.astm.org/BOOKSTORE/.
3. H. Dagnall (1986) *Exploring Surface Texture*, Rank Taylor-Hobson, Leicester, United Kingdom.
4. P. J. Blau (1989) *Wear Transitions of Materials: Break-in, Run-in, Wear-in*, Noyes Publishing, Park Ridge, NJ, pp. 418–426.
5. R. L. Jackson (2012) "Lubrication," in *Handbook of Lubrication and Tribology*, Vol. 2, R. W. Bruce, ed., CRC Press, Boca Raton, FL, pp. 14-1–14-14.
6. P. R. Rubiolo (2006) "Probabilistic prediction of fretting-wear damage of nuclear fuel rods," *Nuclear Engineering and Design*, Vol. 236, pp. 1628–1640. Work rate.
7. P. J. Blau, B. C. Jolly, W. H. Peter, and C. A. Blue (2007) "Tribological investigation of titanium-based materials for brakes," *Wear*, Vol. 263 (7–12), pp. 1202–1211. Metric for break energy.
8. L. C. Buckman (1998) *Commercial Vehicle Braking Systems: Air Brakes, ABS, and Beyond*, SAE International, Vol. 1405, Society of Automotive Engineers, Warrendale, PA.
9. M. Steiner, M. Mitsias, K. Ludwig, and M. Kern (2009) "In vitro evaluation of a mechanical testing chewing simulator," *Dental Materials*, Vol. 25, pp. 494–499.
10. P. Lambrechts, E. Debels, K. Van Landuyt, M. Peumans, and B. Van Meerbeek (2006) "How to simulate wear? Overview of existing methods," *Dental Materials*, Vol. 22, pp. 693–701.

Chapter 6

Wear Problem Solving—
The Next Steps

Earlier chapters in this book defined a tribosystem and what is meant by triboelements. They presented tools to detect wear problems and measure their effects, and they described the characteristics of the common forms of wear and surface damage. Included in those descriptions was a hierarchical approach to organize surface damage based on the type of relative motion, characteristic features, and debris. The differences between a wear type, a wear process, and a wear mechanism were discussed. Importantly, it was noted that different types of wear can occur simultaneously on nearby locations on the same part and that wear is often seen to progress through a series of stages or transitions. Such diverse complications notwithstanding, a tribosystem analysis (TSA) form was presented in Chapter 5 to help organize information about specific wear- and friction-related problems. Completing a TSA form is not an end in itself but rather can lead to one or more approaches to a solution.

Since all relevant attributes of a given tribosystem may not be known, a TSA should be used in conjunction with tribotesting and the opinions of experts who have successfully solved similar kinds of problems in the past. Fortunately, there can be more than one viable solution to a wear problem. All that remains is for the practitioner to find just one of them that works. This chapter explores the overarching and well-known scientific method: using the TSA to define the problem, analyze it, apply a hypothesis, test that hypothesis (validation), iterate on possible solutions, and select the one that works best.

Despite the wide acceptance of the scientific method, it is surprising that so many authors who submit papers to wear-related technical journals seem to have ignored it. After a vague justification for why they are doing the work, they simply state what was done and report the data, perhaps with curve-fitting to generate an empirical model. They often adopt a linear wear model despite the absence of evidence that the wear in that tribosystem is indeed linear with time or sliding distance. Unfortunately, it is deceptively simple to rub materials together in a cobbled-together apparatus of some kind and curve-fit the data. While authors may provide detailed characterization of surface features using sophisticated analytical tools, they fail to provide and test a logical hypothesis for their approach based on sound principles and hard-won previous data in tribology. It is unfortunate to see so much effort placed on characterization when the basic premise or selection of a test method is faulty.

Remarkably, many wear investigators use a pin-on-disk machine to address all sorts of unrelated wear or friction problems. Recently, the author estimated that as many as 12%–17% of tribology conference papers and instrumentation surveys use pin-on-disk machines despite the fact that very few practical engineering wear situations display this kind of geometry [1]. In all likelihood, it is either for convenience (pin-on-disk machines are readily available, they seem to be easy to build, and standards exist for using them) or a lack of the investigator's understanding of the principles of tribosystem analysis that leads to their selection.

This concluding chapter begins with a list of options for addressing wear problems, explains what kinds of information tribotesting can provide, discusses test selection by TSA matching, and concludes by mentioning some pitfalls associated with meeting specifications for wear behavior based on a test method that bears little correlation to the intended application.

6.1 Options for Wear Problem Solving

There are at least a dozen possible avenues for solving friction and wear problems. Not all of them may be viable in specific cases, so the practicality of each needs to be evaluated and then accepted or rejected. Having eliminating the "impractical" options, one can focus on the more promising ones. The 12 options are as follows:

1. *Do nothing—wait and see.* The components might perform acceptably for a while longer. Put off action until the function of the tribosystem is so degraded that the situation can no longer be tolerated.
2. *Replace the worn parts.* Replace worn parts with new parts of the same kind.
3. *Repair or recondition surfaces of worn parts.* Examples include replating and finishing worn bearing surfaces with hard coatings or regrinding railroad rails to restore their profiles and to remove near-surface cracks. Included in this approach is the use of a "wear allowance" that is engineered into the part dimensions with the expectation of future refinishing.
4. *Use monitoring methods to detect incipient failures.* Install sensors or monitoring systems to detect potential or developing wear or friction problems. Some common methods are described in Chapter 2.

5. *Change the severity of operating conditions of the equipment.* Reduce the severity of operating conditions (e.g., tell a driver not to "ride the brakes") or educate the primary users. It can be difficult to implement this approach unless the machine can be operated under less severe operating conditions but still perform its primary function. Trusting the operators to "back off" on pushing their machinery too hard is problematic in an environment where productivity or speed is rewarded.

6. *Replace the parts with a better performing component.* Choose equipment that has a history of fewer friction or wear problems. Perhaps a different make or model of pump or seal or bearing will perform more reliably than the current one.

7. *Redesign the problem equipment.* Redesign the current machinery to minimize the severity of conditions placed on the contacting parts. Examples include substituting tapered bearings for spherical bearings in a gearbox or changing the type and profile of gearing in a power transmission.

8. *Improve installation or running-in procedures to ensure proper alignment and surface conformity before long-term operation.* In precision bearing contacts, a fraction of a degree of misalignment arising from installation errors or cumulative wear could result in stress concentrations or instabilities in operating dynamics that could lead to high stress concentrations and premature failure. Establishing an effective running-in procedure may require some testing or even trial and error. Some tribosurfaces run in naturally and effectively upon start-up but others require proactive procedures [2,3].

9. *Improve the lubrication system.* Improve filtration, monitoring, or the method of delivery to avoid debris accumulation.

10. *Replace the current lubricant with a better performing one.* Sometimes, a previous-generation component may be asked to perform under more demanding conditions when a machine design is upgraded. Two examples might be using a previous transmission with a more powerful engine or scaling up a 650-kW-sized wind turbine gearbox to a 5-MW unit without changing its basic design. A lubricant that may have worked satisfactorily in the past may not be up to new performance demands for higher temperatures or loads. An improved formulation may be required.

11. *Replace the contacting materials, surface treatments, or coatings.* Use an alternative material or surface engineering process designed to perform well in similar applications. This includes using special finishes or patterns on the surfaces as well as changes in compositions and heat treatments for existing materials.

12. *Develop new materials or lubricants tailored specifically for this application.* This is invariably a riskier, costlier, longer-term solution and may require a combination of research and development, screening tests, and field trials.

In practical situations, human factors contribute to wear problems. Operators may force machinery to perform at a higher level than for which it was designed.

Sometimes, lubrication intervals recommended by a manufacturer are ignored (for example, omitting the periodic recommended lubrication of wire ropes, cables, and chains in mining equipment). Better training or supervision of machine operators can help.

Not only are there different academic disciplines and engineering practices involved in implementing the foregoing solutions, but also the associated costs and efforts differ widely. Recall the old adage that if all you have is a hammer, all problems begin to look like a nail. That is, if you are a design engineer, then the solution must involve a redesign, but if you are a lubricants specialist, then the solution is to find a better lubricant formulation. If you are a metallurgist, a new heat treatment or coating is the answer. Exceeding one's comfort level may be required when implementing a truly multidisciplinary approach. Hypothesis testing remains a key tool.

Ideally, sufficient resources would enable multidisciplinary approaches to be pursued in parallel. Such a conservative, albeit costly, effort is based on the reduction of possibilities and should be considered as enlightened down selection rather than trial-and-error (i.e., testing every alloy on the storeroom shelf until, hopefully, one works).

Few businesses have the resources to establish a research and development program or to hire a group of full-time tribologists to solve wear and friction problems. They generally prefer to find something that works, patch the problem, and move on.

6.2 The Purpose for Tribotesting and the Type of Information Such Tests Can Provide

The nature of wear has been studied for hundreds of years, and the origins of friction, much longer [4], yet the complexity of practical wear problems may not permit one to apply a simple equation or a first-principles modeling approach to confidently select materials, surface treatments, or lubricants—especially for new or untried designs. Nevertheless, catalogs for bearings, gears, or seals selection commonly contain technical sections or tables of design parameters to enable engineers to make component selections if one knows the operating conditions. Such empirical relationships may be applicable to properly installed components, and they presume that only one major failure mode occurs (e.g., rolling contact fatigue life that fits a statistical distribution). While helpful, these are based on controlled test rigs and do not apply when other failure modes occur (e.g., overheating, third-body contamination, installation errors). As a result, engineering approaches to wear problem solving usually involves a combination of testing, analysis, and validation. Would the reader feel confident about riding aboard a prototype airplane that has never been flown before and that has been designed based entirely on theory without any component testing? Proper testing is essential.

What can tribotests tell us? Tribotests are intended to characterize the wear and/ or frictional characteristics of materials and lubricants. Numerous surveys, handbooks, and the tribology literature report hundreds of kinds of tribotests. Therefore, the challenge in wear diagnosis and problem solving is to select not only the proper type of laboratory or field test but also the variables to be applied, to determine the repeatability of the data (confidence level) and to analyze their meaning in the context of the specific engineering or fundamental problem being addressed.

The type of information needed should be kept in mind because there is no "universal" wear test. Therefore, the first step in selecting a wear test method or combination of tests is to decide why the data are needed and what metrics would be most valuable. Six reasons for wear testing are as follows:

1. *Fundamental knowledge.* To study the fundamental characteristics associated with a certain type of wear (e.g., basic research into the tribophysics of interfaces, the mechanisms of plowing, the role of adhesion in wear, the nature and properties of asperities, the tribochemistry of lubricant additives, and surface catalysis).
2. *Material or lubricant development.* To enable the development of a new material, coating, surface treatment, or lubricant (either generic or applications directed).
3. *Material selection.* To select or rank a set of materials (or compatible pairs of materials), surface treatments, or lubricants for a specific engineering purpose (component design or improvement of current performance).
4. *Effects of agents.* To investigate the influence of a wear-causing medium when placed in contact with the surface of a material (environmental effects such as tribocorrosion).
5. *Specifications.* To meet commercial specifications (quality control or requirements for product performance, lifetime, durability, and general acceptability for use).
6. *Transitions identification.* To determine critical conditions under which wear transitions can occur. Examples include seizure, scoring, scuffing, galling, lubricant load-carrying capacity, or other critical wear transitions. Some tests are designed to study wear transitions and to indicate their severity if and when they do occur.

In some situations, more than one of these purposes can be involved. Unfortunately, all six kinds of information are not provided by the same test. Some tests may show how a material will fail in wear but not in a way that allows a quantitative metric to be computed or compared. Consider, for example, two cases: how one might quantify the severity of (i) pitting or (ii) galling.

Micropits on rolling contact surfaces form at the asperity level. Their extent can be detected by microscopy or other surface imaging methods (see Chapter 4), and one could decide to measure average pit sizes, spacing, and number of pits per unit area. In other cases, it may be sufficient to know whether the surface is pitted or not. Perhaps, criteria like "minor pitting," "significant pitting," or "severe pitting," along with illustrative images, can be used to differentiate behavior. Figure 3.7 shows a system to describe qualitatively the severity of pitting-related surface damage in rolling contacts. In some types of rolling contact tests, an indirect method, like the onset of a critical level of vibration, is used to terminate the test and establish rolling contact fatigue (RCF) life.

In the case of galling, one could apply a set of conditions and then, using the naked eye, use "yes" or "no" as pass or fail criterion. Other methods such as ASTM galling standards G98 [5] and G196 [6] result in what has been called the threshold stress for galling, but its measurement has been fraught with difficulty because of the lack of repeatability and a dependence upon how the tests are conducted. Sampling becomes an issue because galling may be localized. How the degree of galling is quantified challenges the engineer to go beyond phrases like "slightly galled," "severely galled," or even just "galled." During the 1980s, Ives, Peterson, and Whitenton, at the U.S. National Institute for Standards and Technology, conducted a study of how one might use profiling at specific places on galled coupons to quantify galling and what parameters to use [7]. The number of measurements and where on the surface to sample each time were concerns because a galled feature could lie between standard sampling locations and be undetected if the same procedure were followed each time (e.g., five locations equally spaced along the length of a sliding scar).

If it requires a microscope to find tiny areas that seem to be galled, is it fair to say the surface is galled, or rather is it more correct to call it microwelding, scuffing, or scoring? Once again, terminology plays a role in reporting wear testing results. While design engineers might want the output of a wear test method to yield a single quantitative design parameter (e.g., a wear rate, critical operating envelope, or a not-to-exceed pressure times velocity [PV] limit), some forms of wear testing may not yield an unambiguous numerical result.

The need for a tribosystem analysis is based on the finding that wear rates and the relative wear rankings of materials depend upon the tribosystem characteristics, not only materials' composition, treatment, and resultant basic properties. In other words, *there is no such thing as the inherent wear resistance of a material or pair of contacting materials* because the context of their usage affects their behavior. The same set of material pairs can rank in opposite order of merit in regard to resistance to different forms of wear. For example, Czichos [8] showed that four surface treatments ranked in opposite order or merit in regard to abrasive and adhesive wear resistance. Regardless of the reason for wear testing, therefore, performing even a rudimentary TSA on the primary tribosystem and the tribotest will help to determine the relevance of the data to the problem at hand and aid in the selection of appropriate tests.

There have been ample books, surveys, handbooks, and reviews published on the subject of wear testing. Hundreds of devices are described in the literature. Table 6.1 summarizes the most common wear test geometries and the major forms

Table 6.1 Common Wear Testing Geometries

Test Name	Description (ASTM Designation, If Applicable)	Primary Wear Type[a]
Loop abrasion test	Block pressed against a moving strip of abrasive tape (G174).	2BAb
Pin abrasion test	Pin of test material sliding on a drum or flat platen of an abrasive paper (G132).	2BAb
Taber abrasion index	Counter-rotating miniature grinding wheels on a rotating flat platen (G195).	2BAb
Dry sand/rubber wheel test	Loose silica sand particles held against a rectangular test coupon by a rotating wheel (G65).	3BAb
Jaw crusher impact abrasion	Moving plates rush rocks between them in a severe, high-stress 3-body abrasion situation (G134).	3BAb, RI/2B, RI/3B
Wet sand/steel wheel test	Loose abrasive particles held against a rectangular test coupon by a rotating steel wheel (B611).	3BAb
Wet sand/rubber wheel test	Loose abrasive particles in a slurry held against a rectangular test coupon by a rotating wheel (G105).	3BAb
Slurry abrasivity (Miller number)	Assessment of the abrasiveness of a slurry against one or more reference materials, usually metals (G75).	3BAbr
Block-on-ring sliding wear	Flat rectangular block sliding against a rotating ring, a pin or conformal "shoe" is also used (G77, D2714).	S/NU
Pin-on-disk sliding wear	Pin specimen sliding against a rotating disk. Some configurations use multiple pins, others a single pin (G99).	S/NU
Sliding face seal	Annular ring of material sliding against a similar face surface in a face seal geometry.	S/NU
Reciprocating pin-on-flat	Pin specimen sliding back and forth against a flat specimen surface (G133).	S/NR

(Continued)

Table 6.1 (Continued)   Common Wear Testing Geometries

Test Name	Description (ASTM Designation, If Applicable)	Primary Wear Type[a]
Button-on-block	Either unidirectional or oscillating face-on-face sliding, commonly used to assess galling (G98, G196).	S/NU or S/NR
Multidirectional ball-in-socket	Ball-in-socket with variable loads, speeds, and angular motions such as those that can simulate human hip joints.	S/MD
Fretting contact	Low-amplitude, oscillatory contact (G206). Typical fretting tests use linearly reciprocating motions of a ball or pin against a flat surface. Other types of fretting, like pivoting or rocking, generally require custom apparatus.	S/NR-Fr
Solid particle impingement erosion	Gas-entrained erodant particle stream aimed at various angles at a flat coupon (G76).	Er-SIm
Liquid droplet erosion	Atomized steam of liquid droplets aimed at various angles at a flat surface.	Er-LDr
Liquid jet erosion	Liquid jet impinging on a surface at various angles in air.	Er/Jet-O
Cavitation erosion	Oscillating cone producing a bubble field in a gap and near the surface of a test specimen (G32).	Er-Cav
Cavitating jet erosion	Submerged jet of cavitating fluid (G134).	Er/Jet-S
Spark erosion	Electrical arc made to bridge from an electrode to the specimen surface.	Er/Spk
Ring-on-ring	Rotating rings turning on parallel axes and with separately adjustable speeds to produce variable slip and rolling conditions.	RCF-Slp

(Continued)

Table 6.1 (Continued)   Common Wear Testing Geometries

Test Name	Description (ASTM Designation, If Applicable)	Primary Wear Type[a]
Captive cylindrical button	Three rollers spinning on a captive cylindrical surface (button), used for micropitting or rolling contact fatigue.	RCF-Slp
Balls-on-rod	Three balls clamped against a spinning rod, used for rolling contact fatigue.	RCF-P or RCF-Slp
Inclined sphere-on-disk	Designed for lubricant screening; uses a ball rotating on an axis inclined to the plane of a rotating horizontal disk. Slide/roll ratios can be varied.	RCF-P or RCF-Slp
Hertzian contact fatigue	Vertical oscillation of a sphere or other simple shape.	RCF-VC
Four-ball lubricant tests	Popular method of lubricant screening using a spinning ball resting on a nest of three captive lower balls of the same diameter. The diameter of the wear scar is used as a measure of lubricant performance. Such tests generally use standard steel as the ball material, so this is more a test of lubricants than it is of bearing materials.	S/NU

[a] See Chapter 3 and Figure 3.3.

of wear they address (based on the category codes in Figure 3.3). Some tribotests have been standardized and others have not. Despite the range of geometries and testing conditions represented in Table 6.1, not all wear situations can be replicated by those tests, and customized test rigs may still need to be developed to better simulate the conditions identified in a TSA.

6.3 TSA Matching—Field versus Laboratory

Accomplishing a TSA form summarizes what is known about the operating conditions and modes of wear failures. It also reveals what is unknown about them. Still, a TSA can also help when selecting appropriate test methods. Tribotesting can be used along with any or all of the six options listed in Section 6.1. Since a tribotesting apparatus is also a tribosystem, matching as many characteristics of the testing machine's TSA with those in the primary tribosystem of interest increases the likelihood that the test data will be relevant for solving specific wear problems. It is particularly important to match the type of surface damage produced on the materials even if not all aspects of the two tribosystems can be matched.

Figure 6.1 depicts the matching process schematically. Not only matching the TSAs is shown, but the use of relevant supporting data from other kinds of tests is also depicted. Like the TSA match, the conditions used in supporting tests, shown below the dashed line in the figure, should be viewed in terms of how well they simulate operating conditions, or at least correlate with them. For example, Rockwell hardness may correlate with the relative performance of a set of heat-treated gear steels even though the slow indentation of a carbide ball is quite different from the rapid engagement and disengagement of mating counterformal gear teeth. It is incumbent upon the investigator to determine the validity of such correlations especially when the supporting test method does not match the field application very well.

In another example, the so-called four-ball test (i.e., three captive lower balls and a spinning upper ball in a tetrahedral arrangement and immersed in a lubricant) has been used for many years in the oil industry. Sometimes it is used to screen oils, and sometimes, to screen greases [9]. The convenience of the method and its familiarity in the lubricants industry are two things in its favor; however,

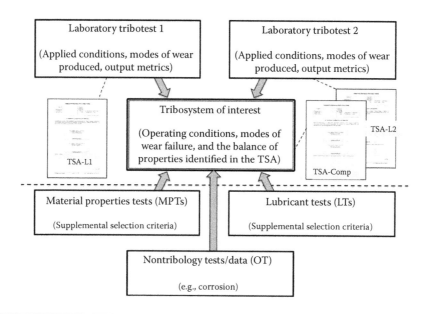

Figure 6.1 Matching the TSA of the test method to that of an application. Supplementary data from other sources are shown below the dashed horizontal line.

it is difficult to find any practical application involving this kind of contact geometry. One therefore needs to look closer to see if, for example, the applied speeds, contact pressures, surface finishes, and temperatures relate to the application. The alternative is to establish such a correlation through a series of field trials and laboratory trials using the same lubricants and bearing materials. Bracketing the range of responses using good and bad performing materials or lubricants can be helpful to benchmark such tests.

As noted earlier, in applications-oriented testing, the type of wear produced should replicate that type of wear that occurs in the application. Side-by-side comparisons of wear surfaces produced in the field and laboratory should display similar features. If a different form of wear is produced by a laboratory test, then its value in material selection becomes questionable because the material is wearing by different processes in the test and in the field.

As Figure 6.1 also shows, below the dashed line, other types of material or lubricant characterizations may be useful. Those kinds of characterizations may not lend themselves to a tribosystem analysis but can provide essential ancillary inputs for decision making. For example, hardness tests, corrosion tests, fracture toughness tests, fatigue tests, thermophysical property measurements, and lubricant property tests may be valuable supplements to wear tests.

A caution about accelerated testing. Attempts to "accelerate" the speed of producing data by raising the normal load or increasing speed so that measurable amounts of wear can be generated in a convenient period potentially can produce results that do not simulate the normal wear response of the targeted tribosystem very well. If a material in a given application wears by mild abrasion, then why would a severe abrasion test bear relevance to the problem? Comparing TSAs for a candidate test apparatus and the field component can highlight differences that will help the engineer to decide whether that test is appropriate.

Not only the state of the as-worn surfaces but also the characteristics of the debris particles can indicate whether the laboratory test is appropriately simulating the form of wear of interest. The identification of wear types based on such techniques as ferrography (see Section 2.5.4) can also be used to analyze laboratory wear test results for relevance to field applications. The morphological characteristics of wear particles—flakes, chips, fine dust, oxidized, metallic, and so on—can secondarily confirm that a laboratory test replicates the application.

When a new machine design is in the conceptual or prototype stage, there will likely be no samples of field-worn specimens available to examine. In this case, there are two possible approaches: (a) benchmark the wear requirements with those from an existing machine that operates under conditions similar to those in the new design and (b) perform a TSA and, using what is already known about wear in the behavior of materials under those conditions, estimate the most likely form of wear that would occur.

The type of wear expected for a new, unproven machine design may be straightforward to envision, or it may be unknown. For example, geothermal drilling is likely to occur in a hot environment of slurry erosion, possibly with 2-body abrasion and impact. This complex environment is challenging to simulate in the laboratory, and existing wear test methods, like a dry sliding pin-on-disk tester or a Taber Abraser™ (in essence, a load-controlled, 2-body rotary grinding test), may fall short of providing sufficient simulation to use for material selection in geothermal drilling components. Their TSAs are not similar enough.

When designing a new type of rolling element bearing, rolling contact, possibly with some degree of slip, is expected and bearing testers of many types are already available to simulate the expected loads, speeds, and slip ratios expected for such new designs. However, when determining the propensity for micropitting of a bearing surface, the appropriate slip ratio should be chosen to produce the kind of pitting observed in field components. Too much slip, and the failure mode may begin to resemble sliding wear, scoring, or scuffing rather than micropitting. Any pits that form can be distorted or smeared over if the slip ratio is too high. On the other hand, the damage form must simulate that for the application.

In summary, the dilemma faced by materials and design engineers is how to select materials or lubricants for an untried application without the resources or time to build an expensive simulator that may or may not turn out to simulate the service environment sufficiently well to produce useful data. It may be necessary to compromise in such cases, using an available test method that is "close enough" to the conditions in the application's TSA, to provide at least first-order screening of a set of candidate materials or lubricants. The leading candidates from the down-selected group can then be subjected to field trials. Recall the hierarchy of wear testing component size scales embodied in the German standard DIN 50 322 [10].

Three case studies to illustrate attempts to match field performance to laboratory tests follow. The first is a friction-related case, and the next two involve wear.

6.3.1 Case Study 1: Friction Coefficient versus Field Performance for Brake Lining Materials

In the truck brakes industry, in the early 2000s, it can cost upward of US$10,000 to conduct a standard, weeklong, multistage wear test of a new friction material (lining/disc or lining/drum) combination. Full-scale inertia dynamometers and smaller systems, like a wear-modified version of the friction assessment and screening test (FAST), the subscale Chase test for a block-on-brake drum [11,12], or a custom-built, bench-scale tribometer [3], could, in principle, be used to prescreen a large set of friction materials in a shorter period and at lower cost. Yet basing a final material selection or product commercialization decision on bench-scale test results is risky, even when reasonable correlations to material performance rankings have been done with full-sized components. In the field of friction brake materials development, one of the most difficult aspects of simulative testing is driver behavior. Issues like regional geography, weather, and service type (short-haul delivery, on-highway, commercial transit, ambulances, etc.) directly affect wear life, especially if it is measured not in pad thickness per unit distance slid but rather by the more common metric of vehicle miles driven between pad replacements.

A study comparing original equipment truck brake linings versus aftermarket linings, sponsored by the U.S. Department of Transportation, was performed in the early 2000s at Oak Ridge National Laboratory (ORNL) [13]. Part of that investigation involved a correlation of on-highway and proving ground dump truck braking data with friction data from a commercial test laboratory (the Chase test) and from a laboratory-scale tribometer built at ORNL that was called the subscale brake materials tester (SSBT) [12,14].

Four original equipment linings and four aftermarket replacement linings were used. The commercial proving ground was the Transportation Research Center, Inc. (TRC) in East Liberty, Ohio. Braking tests done there conformed to U.S. Federal Motor Vehicle Safety Standard FMVSS-121 [15]. Using the same trucks and linings, tests were then subjected to braking tests on a stretch of highway near Morristown, Tennessee, using drivers from local Knox County hauling companies and under supervision by the Tennessee Highway Patrol. Standard Chase tests (SAE J 661) [11] were performed at a commercial testing laboratory in Detroit using samples of the same lining materials. Chase tests involved a 25 × 25 mm square coupon of the brake lining pressed against the inside of a rotating cast iron drum under both cold and hot conditions. SSBT tests used the apparatus shown in Figure 6.2. They were performed under two different pressure × velocity (PV) conditions after applying a burnishing process to condition the disc surfaces: high PV = 300 N × 15 m/s (or 4500 J/s) and low PV = 167 N × 6 m/s (or 1000 J/s).

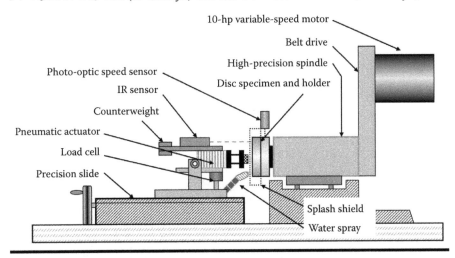

Figure 6.2 The subscale brake materials testing apparatus (SSBT) built at Oak Ridge National Laboratory consists of a 125-mm-diameter disk and a 12.7 × 12.7 square pad pressed against it using a hydraulic actuator. Friction force, wear, and disc temperature (using an infrared sensor) can be measured. Scaled highway speeds were based on the ratio of vehicle brake disc/tire diameter.

Reducing the data from the different types of tests for a fair comparison proved to be challenging in light of the different sets of variables present in the three tribosystems: (i) dump truck stopping on a road surface, (ii) truck stopping under controlled conditions at a proving ground (different driver), (iii) Chase tests run at two temperatures, and (iv) SSBT tests run at two PV levels. The six-stop average stopping distances for a truck at the TRC showed stopping advantages of original equipment linings over aftermarket linings (see Figure 6.3). Within the repeatability of the methods, the four original equipment linings stopped the vehicle faster than the aftermarket linings.

Figure 6.4 compares friction coefficients measured in two types of laboratory tests with the average stopping distances obtained for the two types of linings under proving ground conditions. There was a significant difference in SSBT data for low PV (upper solid circles) and for high PV (bottom set of solid circles). The two types of tests that best discriminated between the original equipment linings and aftermarket linings were the Hot Chase test and the Low PV SSBT test. In summary, the normal Chase test data and the High PV SSBT data showed little effect of lining type on friction coefficient in this case. Like the trend in stopping distance in Figure 6.3, shorter distances of original equipment linings correlated to increased friction coefficients, but only in the Hot Chase and SSBT Low PV cases. This case study demonstrates that obtaining laboratory–field correlations when matching TSAs is a function not only of picking the appropriate laboratory tribosystem but also in selecting the correct set of variables to be applied when running tribotests on that system. One set of test variables may differentiate material performance better than another.

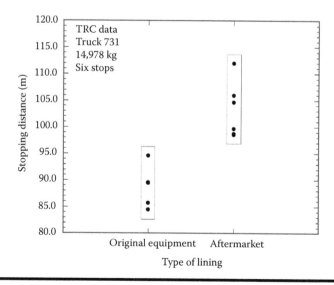

Figure 6.3 **Transportation Research Center (TRC) data on stopping the same truck with different linings under controlled test conditions.**

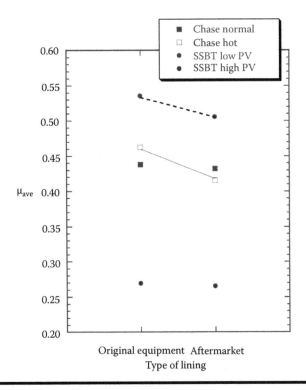

Figure 6.4 **Comparison of friction coefficients measured in laboratory tests with truck stopping distances.**

6.3.2 Case Study 2: Spectrum Loading Tribotests to Simulate Engine Bearings

The forces on a large-end bearing on a connecting rod in a fired internal combustion engine vary with the angle of rotation of the crankshaft. A tribotester was designed to apply a programmable varying test load to a simple shaft-on-flat coupon geometry with drip-fed lubrication (see Figure 6.5). The spectrum of applied loads was based on data published on the relative variation of forces in engine connecting rod bearings during rotation. As the forces change during rotation, so does the regime of lubrication and film thickness. Therefore, wear can occur intermittently if the film thickness is insufficient to separate contact surfaces during portions of rotation. This methodology was used to compare the friction coefficient and wear behavior of candidate lightweight materials and surface treatments as candidates for large-end bearings [15].

An interesting preliminary set of tests within the cited study [16] compared the effects of variable loads with constant loads using the same speed and test duration. The test couple was Ti-6Al-V4 alloy (flat coupon) sliding against a rotating cylinder of crankshaft-type alloy steel (AISI 8620) under drip lubrication by 15W40

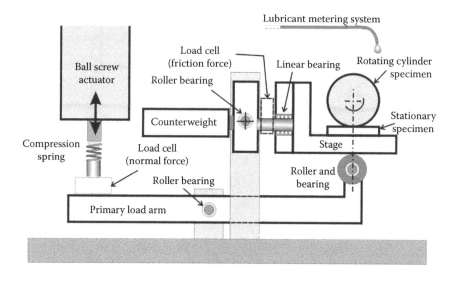

Figure 6.5 Variable load-bearing test system developed at Oak Ridge National Laboratory and used in Blau et al. [16]. The roller at the lower right pushes the flat specimen up against a rotating cylinder. The actuator drive is controlled by a computer to apply a spectrum of loads.

diesel engine oil (Valvoline "Blue"). One experiment was run using a constant load of 50 N and the other test used a spectrum of loads ranging between 10 and 95 N, whose average load was also 50 N. Results are summarized in Table 6.2. It could be argued that while the average loads were the same in both cases, the higher load portion of the loading spectrum contributed to the 65% higher wear volume, or in this case an additional 0.0126 mm³ in wear volume over the length of the experiment.

Table 6.2   Comparison of Wear Volume of a Lubricated Flat Specimen of Ti-6Al-4V against AISI 8620 Steel under Constant and Spectrum Loads

Sliding Velocity (m/s)	Applied Normal Load (N)	Test Duration (s)	Wear Volume (mm³)
0.50	50 (constant)	1100	0.0193
0.50	10–95 (simulated connecting rod spectrum); average = 50	1100	0.0319

6.3.3 Variable-Condition versus Constant-Condition Tribotests

With notable exceptions, like brake industry test protocols for friction materials described earlier, step-loading tests for load carrying capability, and spectrum-load tribotests like that described in Section 6.3.2, most sliding wear tests used in materials selection or in laboratory-scale wear research and development usually consist of applying constant load, speed, and duration. While convenient, more easily controllable, and easier to benchmark, constant-condition wear tests do not necessarily replicate the varying conditions experienced in many types of practical engineering systems. An additional example of this is the act of starting a cold internal combustion engine. This involves transient heating, changes in lubricant properties, potential scuffing before full oil films develop, and changes in engine rpm. Components like piston rings rubbing on cylinder bores, large and small-end connecting rod bearings, crank shafts, valve guides, fuel injector needles, and bucket lifters can experience a variety of contact conditions sometimes over a short period (many cycles per second). How then can one use a simple unidirectional sliding test, like a pin-on-disk, to simulate such varying conditions? The sobering answer is that, often, one cannot use a simple, constant-condition test to simulate many common engineering systems, and the process of performing a tribosystem analysis reveals such shortcomings.

6.4 Wear Testing to Meet Specifications— Some Potential Pitfalls

A few test methods are engrained in engineering technology and have attained a measure of popularity but are sometimes incorrectly applied to practical problems that lie outside the scope for which these tests were originally intended. The so-called friction assessment and screening test (FAST) method was originally developed as a means to quickly screen candidate brake pad compositions without conducting more expensive full-scale inertia brake dynamometer tests. While engineers who developed the FAST procedure recognized its technical limitations, the test method's simplicity, relatively low cost, and output of a series of figures of merit for frictional performance led to its spread within the friction materials industry. Brake specialists in the industry recognized that the FAST did not simulate a number of key characteristics required to screen materials for full-sized brakes, but their voices were unheard as the popularity of the coding scheme spread [17]. So convenient was the coding scheme that some purchasing agents for public or commercial vehicle fleets adopted this technically flawed method to specify and purchase replacement brake linings.

Another example of potential misuse is the Taber abraser test used to evaluate hard platings for certain applications that do not produce the same wear types or degrees of surface damage [18]. This 2-body abrasion test (in which debris is vacuumed from the contact surface during the test) involves a pair of dual grinding wheels rotating on a horizontal axis against a rotating flat plate. Typically, the wear loss (Δm, in milligrams) is normalized by the number of rotations of the lower plate to provide a so-called Taber index (TI); thus,

$$TI = \Delta m/1000. \qquad (6.1)$$

The Taber index is commonly used to benchmark hard electroplates for their wear resistance. The Taber indices for hard Cr plate, for example, are typically in the range of 2–5 mg/1000 cycles. Suppliers of hard coatings, irrespective of the products' end application, may be required by their customers to meet industry standard TI specifications that do not simulate the end application as revealed by a TSA. Say, for example, that an application concerns a harsh slurry erosion environment like drilling oil wells. Even without a detailed TSA, it should be evident to the reader at this point that such a rotary grinder type test does not simulate certain key aspects of these tribocorrosion-related applications:

- The abraser is a dry, 2-body abrasive wear test in which debris is removed from the wear track by an air nozzle. Its results are dependent on the grade of grinding wheels selected, the manner in which they are periodically dressed, the applied load, and the number of cycles applied between successive measurements.
- A down-hole drill sees wet slurry conditions with third bodies caught in the interface. It can also involve crushing and impact conditions (high-stress impact abrasion). Neither chemical nor third-body erosive mechanical effects are simulated by the Taber abraser test.

From the standpoint of TSA matching, it makes little sense why the data from a popular test method should be used to screen the acceptability of materials for an application whose conditions it does not simulate unless a rigorous correlation has been conducted to validate the method's relevance for that particular application. More likely, a test such as ASTM G105 [19] or even ASTM B611 [20] is more relevant to the stated application. While the latter methods also do not involve impacts, at least they involve wear in a wet, 3-body abrasive environment. The alternative is to design and build a simulative testing system and attempt to establish a specification for its use and applicability or conduct tests on the full-sized machine.

Based on a long, productive life and career in tribology consulting and education, Professor Ken Ludema asserted the following in his textbook [21]:

> Experience shows time after time that simple wear tests complicate the prediction of product life. The problem is correlation or assurance of simulation. For example, automotive company engineers have

repeatedly found that engines on dynamometers must be run in a completely different unpredictable manner to achieve the same type of wear as seen in engines on cars in suburban use. Engines turned by electric motors, though heated, wear very differently from fired engines.

A standard laboratory test method may be conveniently available, and it may provide metrics that can be compared to target values to determine "acceptability," but unless a correlation of that laboratory test with the performance of the intended application has been conducted, such results could turn out to be misleading and costly. Tribosystem analysis, when properly applied, especially in conjunction with intelligent tribotesting, can avoid missteps on the path to successful wear problem solutions.

References

1. P. J. Blau (2014) "The use and mis-use of the pin on disk wear test," presented at the STLE Annual Meeting, May 21, Orlando, FL.
2. P. J. Blau (1991) "Running-in: Art or engineering," *Journal of Materials Engineering*, Vol. 13, p. 47.
3. P. J. Blau (1989) *Friction and Wear Transitions of Materials: Break-in, Run-in, Wear-in*, Noyes Publishing, Park Ridge, NJ.
4. D. Dowson (1998) *History of Tribology*, 2nd ed., Professional Engineering Publishing, London, United Kingdom.
5. ASTM G98-09 (2014) "Standard test method for galling of materials," in *ASTM Annual Book of Standards*, Vol. 03.02, ASTM International, W. Conshohocken, PA, pp. 417–419.
6. ASTM G196-08 (2014) "Standard test method for galling resistance of material couples," in *ASTM Annual Book of Standards*, Vol. 03.02, ASTM International, W. Conshohocken, PA, pp. 872–877.
7. L. K. Ives, M. B. Peters, and E. P. Whitenton (1989) *The Mechanism, Measurement, and Influence of Properties on the Galling of Metals*, NIST IR 89-4064, Final Report, National Institute of Standards and Technology, Gaithersburg, MD.
8. H. Czichos (1978) *Tribology—A Systems Approach to the Science and Technology of Friction, Lubrication, and Wear*, Elsevier, Amsterdam, the Netherlands.
9. ASTM D2266-01 (2015) "Standard test method for wear preventative characteristics of lubricating grease (four-ball method)," in *ASTM Annual Book of Standards*, Vol. 5.01, ASTM International, W. Conshohocken, PA, http://www.astm.org/Standards/D2266.htm.
10. DIN 50 322 (1986) *Kategorien der Verschleißprüfung (Categories of Wear Testing)*, BAM, Berlin, Germany.
11. SAE J661 (2012) *Recommended Practice for Brake Lining Quality Test Procedure*, SAE International, Warrendale, PA.
12. P. J. Blau (2001) *Compositions, Functions, and Testing of Friction Brake Materials and Their Additives*, Technical Report, ORNL/TM-2001/64, Section 5.2, Oak Ridge National Laboratory, Oak Ridge, TN, p. 19, http://www.osti.gov/bridge.

13. W. Knee (project manager) with G. Capps, O. Franzese, P. Blau, D. Rice, B. Jolly, J. Joseph, R. Landis, and D. Boshears (2004–2006) "Heavy truck brake lining performance characterization: Original equipment vs. aftermarket linings," http://web .ornl.gov/sci/engineering_science_technology/technical_articles_for_public_site /Heavy%20Truck%20Brake/heavytruck.shtml.

14. P. J. Blau, B. C. Jolly, W. H. Peter, and C. A. Blue (2007) "Tribological investigation of titanium-based materials for brakes," *Wear*, Vol. 263 (7–12), pp. 1202–1211.

15. Federal Motor Vehicle Safety Standard (2005) *Laboratory Test Procedure, FMVSS-121V-05, Air Brake Systems*, U.S. Department of Transportation, National Highway Safety Administration, Office of Vehicle Safety Compliance, Washington DC, 83 pp.

16. P. J. Blau, K. M. Cooley, and D. Bansal (2013) "Spectrum loading effects on the running-in of lubricated bronze and surface-treated titanium against alloy steel" *Wear*, Vol. 302 (1–2), pp. 1064–1072.

17. SAE 866 (2012) *Friction Coefficient Identification and Environmental Marking System for Brake Linings*, SAE International, Warrendale, PA.

18. ASTM G195-13a (2014) "Standard guide for conducting wear tests using a rotary platform abraser," in *ASTM Annual Book of Standards*, Vol. 03.02, ASTM International, W. Conshohocken, PA, pp. 864–871.

19. ASTM G105-02 (2014) "Standard test method for conducting wet sand/rubber wheel abrasion tests," in *ASTM Annual Book of Standards*, Vol. 03.02, ASTM International, W. Conshohocken, PA, pp. 450–458.

20. ASTM B611-13 (2014) "Standard test method for determining the high stress abrasion resistance of hard materials," in *ASTM Annual Book of Standards*, Vol. 03.02, ASTM International, W. Conshohocken, PA, pp. 13–18.

21. K. C. Ludema (1996) *Friction, Wear Lubrication—A Textbook in Tribology*, CRC Press/Taylor & Francis, Boca Raton, FL, p. 189.

Index

This index includes the preface. Page numbers with f and t refer to figures and tables, respectively.

A

Ablation, 33
Abnormal wear, 84t
Abradants, 41, 42f, 43, 89t
Abrasions (S/Ab), 41–51
  abradants, 41, 42f
  ASM classification, 32, 33
  combined (S/2-3Ab), 38f, 51, 51f
  debris, 37, 41, 46, 48f, 86t
  definition, 41, 89t
  ferrogram particles, 23, 24f
  gerotor pumps, 152
  grinding, 93t
  nomenclature families, 88t
  occurrence in industry, 39t
  organizational diagram, 38f
  3-body abrasions (3ABAb), 47, 49–51, 97t;
    see also 3-body abrasions
  2-body abrasions (2ABAb), 41–47; 2-body
    abrasions
Abrasive, 89t
Abusive wear, 84t
Accelerometers, 13
Acoustic emission (AE), 9–10, 9f, 11t, 12t, 12–15
Additives' functions, resources for analysis of
    wear problems, 99
Adhesive transfer, 52f, 77f
Adhesive wear; *see also* Nonabrasions
  definition, 38, 52, 89t
  ferrogram particles, 23–24, 24f
  human teeth and, 154
  occurrence in industry, 39t
  organizational diagram, 38f

AE, *see* Acoustic emission
AEM, *see* Analytical electron microscopy
AES, *see* Auger electron microscopy
AFM, *see* Atomic force microscope
Aguilar, D. A., 14
Aircraft windshields, 66
Airport luggage carousel, 50, 50f
AISI 304 stainless steel, 50
AISI 440XH stainless steel, 73
AISI 8620 steel, 171, 172t
AISI 52100 steel, 47f, 48f, 70
Aluminum alloy, 151
Analytical electron microscopy (AEM), 106f
Andersson, P., 71
Anvil, 61, 62f
Archard, J. F., 54
Asperities, 71, 77f, 87, 89t
ASTM B611, 174
ASTM D1404, 26
ASTM D7899, 17
ASTM G02.30, 38–39
ASTM G02.40, 39
ASTM G40, 66
ASTM G40-13b
  abrasive wear, 41
  erodant, 64
  micropitting, 71
  significance, 39, 87
  tribocorrosion, 78
  tribolement, 137
  wear, 36
ASTM G73-10, 66
ASTM G75-07, 67
ASTM G98, 162

ASTM G105, 174
ASTM G105-02, 67
ASTM G196, 162
Attachment and exhibits, 130t, 131, 136
Atomic force microscope (AFM), 121t
Attritious wear, 89t
Auger electron microscopy (AES), 106f, 123
Automotive tires, 10, 12t
Axial overloads, 73

B

Babbitt particles, 25
Ball mill balls, 11t
Barrett, Michael, 25
Baruch, Bernard, xii
Bayer, Raymond G., 2, 87
Beach marks, 89t
Bearing failures, 93t
Bearing quality, 13
Bearings
    brinelling and, 58
    ferrograms, 24, 25f
    heat checking, 33
    micropitting, 76, 77f
    plastic, 146
    polishing wear, 49
    roller deterioration, 14t
    rolling contact damage, 73f
    rolling element, *see* Rolling element
        bearings
    significance, 13, 32
    sliding wear, 53–54, 54f
    surface distress, 97t
    2-body abrasions, 45, 47f
Benabdallah, H. S., 14
Bhushan, Bharat, 116
Blau, Peter J., 68f, 85, 87, 107f, 115, 147
Block-on-ring sliding wear, 163t
Blotter spot test, 19t
Boulder dam hydroelectric plant, 68f
Boundary lubrication, 142
Bowden, Frank Philip, 52, 139
Bowen, E. R., 22
Brake
    case studies, 168–171
    discs, 151
    pads, 149
    wear indications, 11t
Break-in, 84t
Bright-field (BF) illumination, 117t
Bright field (BF) microscopy, 106f

Brinelling
    definition, 89t
    false, 32, 58, 91t
    true, 32, 58
    wear indications, 9f
Budinski, K. G., 41
Burning, 33, 89t
Burnishing, 88f, 89t
Bushings, 3, 54, 82t, 146
Button-on-block, 164t

C

Captive cylindrical button, 165t
Case studies
    acoustic emission (AE), 13–14
    brake lining materials, 168–171
    diesel engine valve/valve seat wear, 151
    fuel injector plungers, 152–153
    gerotor cases for fluid pumps, 152
    human teeth, 153–154, 154f
    lightweight rotors, 151–152
    piston ring groove wear, 150
    spectrum loading tribotests, 171–172
    vanes for diesel engine turbochargers, 150–151
    vibration monitoring (VM), 13–14
    wind turbine gearbox bearings, 153
Catastrophic failure, 84t
Cavitating jet erosion, 164t
Cavitation, 67, 90t
Cavitation erosion (Er-Cav), 38f, 67, 68f, 90t,
        98t, 164t
CF, *see* Contact fatigue
Chafing, 90t
Chase tests, 168, 169, 170, 171f
Chatter marks, 90t
Checking, *see* Craze cracking; Heat checking
Chemical wear, 39t, 90t; *see also* Tribochemical
        wear; Tribocorrosion
Chemomechanical planarization (CMP), 50–51
Chemomechanical polishing wear (CMPol),
        38f, 49–51, 78
Chen, X., 58
Chevrons, 75f, 86t, 90t
Chipping, 35, 90t
Closed tribosystems, 4
CMPol, *see* Chemomechanical polishing wear
Cocoa, 90t
Cold starts, 84t, 131, 144, 149
Columbia River, 68f
Combined abrasions (S/2-3Ab), 38f, 51, 51f
Combustion engines, 12t

Complex tribosystems, 2
Condition monitoring, 10
Confocal microscopy, 106f, 118t, 123
Consistent terminology, xiii, 32–33
Contact fatigue (CF), 69–78; *see also* Rolling
 contact fatigue
  definition, 90t
  ferrogram particles, 23–24, 24f
  normal vibrational contact (VC), 76–78; *see
   also* Hertzian contact fatigue
  organizational diagram, 34f, 38f
  pure rolling contact fatigue (RCF-P), 69–73;
   see also Pure rolling contact fatigue
  rolling contact fatigue with slip (RCF-Slp),
   73–76; *see also* Rolling contact fatigue
   with slip
Contact stress, 140–141
Contaminants, 13, 25, 135
Copper particles, 25
Corrosion
  chemical reaction and, 33
  erosion–, 32, 69
  fretting, 35, 92t
  significance, xi, 8, 13
  tribocorrosion, 78–80
   definition, 34, 78, 98t
   dental erosion, 79
   MAC approach, 87
   organizational diagram, 38f
   sequence, 79t
   slurry and, 67
  TSA form, 135, 144
Corrosive wear, 32, 33, 90t
Cowan, Richard S., 13
Cracking damage, 13, 43, 62
Crackle test, 19t
Crater wear, 90t
Craze cracking, 91t
Cross-sectioning, 112–113, 112f
Currency, 62
Cutting
  chip-like debris, 37, 41, 46, 48f, 86t, 111f
  gouging abrasions and, 43
  tools, 11t
  wear, 41, 88f, 90t, 95t; *see also* Abrasions
Cyclic wear behavior, 84t
Czichos, H., 162

D

Dark field (DF) microscopy, 106f
Dark-field illumination, 117t

Davis, J. R., 57
Debris
  abrasions, 37, 41, 46, 48f, 86t, 111f
  adhesive, 86t, 89t
  cocoa, 90t
  cutting chip-like, 37, 41, 46, 48f, 86t, 111f
  delamination, 91t
  fine powdery oxidized, 86t
  flat, shiny flakes, 86t
  fretting, 60, 87, 96t
  metallic flakes, 16, 111f
  particle characterization, 18, 21f, 22
  pristine wear, 55
  resources, 99
  TSA form, 135, 144
  unidirectional sliding, 53, 54f
Debris particle characterization, 18, 21f, 22
Def Stan 05-50 (Part 39) method, 26
Degradation of function, 9, 9f
Delamination, 57f, 91t
Delamination surface layers, 86t
The delamination theory of wear, 55, 56f
Dental erosion, 79, 154
Depth of field (DOF) microscopy, 106f
Destructive pitting, 33
Destructive wear, 32
Diagnostic methods/tools
  acoustic emission (AE), 9–10, 9f, 11t, 12t,
   12–15
  human eye, 104
  light and electron optical imaging tools,
   106f
  light optical microscopy (LOM), 106–116
  most commonly used, 103
  motor current signature analysis (MCSA), 15
  oil analysis (OA), 15–26
  radionuclide wear detection, 27
  scanning acoustic microscopy (SAcM),
   124–125, 125f, 126f, 127
  scanning electron microscopy (SEM),
   123–124
  sketching, 104, 104f, 105f
  stylus profiling and noncontact imaging,
   116, 118–123
  surface imaging, 10, 12
  thermography and thermal wave imaging,
   127
  transmission electron microscopy (TEM),
   106f, 123–124
  vibration monitoring (VM), 9–10, 9f, 11t,
   12t, 12–15
  visual inspection, 10, 12, 103

Diamond turning tools, 12t
Diamond wafer saws, 113
Diesel engine valve/valve seat wear, 151
Differential interference contrast (DIC)
    illumination, 118t
    microscopy, 106f
Diffusion, 33
Diffusional wear, 91t
Digs, curved, 86t
Dimensional change, 9, 9f
DIN 51813 method, 26
DIN 51825 method, 26
Dinolite™ instrument, 108f
Direct reading, 23
Disc plows, 51, 51f
Dodge TDT 825 reduction gearbox, 24
Doi, T. K., 51
Dry fretting, 60
Dry sand/rubber wheel test, 41, 42f, 163t
Dusting, 91t

E

Edge rounding, 113
EDX (energy dispersive X-ray spectroscopy), 106f
Effects, cost, and economics, 62
Electrical burns, 92t
Electrical grounding problems, 8
Electrical pitting, 32
Electroless plating, 113
Electron microprobe analysis (EMP), 106f
Elemental spectroscopy, 19t
Embedded particles, 86t
Embedded wear indicators, 10
Embedment, 88f
EMP (electron microprobe analysis), 106f
Energy dispersive X-ray spectroscopy (EDX),
    106f
Engle, P. A., 62
Environmental scanning electron microscopy
    (Env-SEM), 106f
Env-SEM (Environmental scanning electron
    microscopy), 106f
Erodants, 64, 91t
Er-Cav, *see* Cavitation erosion
Er/Jet-S, *see* Submerged liquid jet erosion
Er-Jet-O, *see* Open liquid jet erosion
Er-LDr, *see* Liquid droplet erosion
Erosion
    −corrosion, 32, 69
    definition, 91t
    erodants, 64, 91t

    extensity of, 64
    factors, 65
    rain, 95t
    wear, *see* Erosion wear
Erosive, multibody (Er)
    ASM wear classification, 32
    occurrence in industry, 39t
    cavitation erosion (Er-Cav), 67, 68f; *see also*
        Cavitation erosion
    erodants, 64
    erosion factors, 65
    extensity of erosion, 64
    liquid droplet erosion (Er), 66; *see also*
        Liquid droplet erosion
    open liquid jet erosion (Er/Jet-O), 68; *see also*
        Open liquid jet erosion
    organizational diagram, 38f
    rain erosion, 95t
    slurry erosion, 66–67, 97t; *see also* Slurry
        erosion
    solid impingement erosion (Er-Slm), 66;
        see also Solid impingement erosion
    spark erosion (Er-Spk), 67; *see also* Spark
        erosion
    steam erosion (Er-Stm), 69f, 69; *see also*
        Steam erosion
    submerged liquid jet erosion (Er/Jet-S), 68;
        see also Submerged liquid jet erosion
Er-Slm, *see* Solid impingement erosion
Er-Slu, *see* Slurry erosion
Er-Spk, *see* Spark erosion
Er-Stm, *see* Steam erosion
Extended depth of focus processing, 107
Eyre, T. S., 39, 39t

F

15W40 diesel engine oil, 172
Face seals, 2t, 91t, 93t, 129, 146, 163t
Faceting, 88f
Failure analysis and prevention (1975), 32–33
False brinelling, 32, 58, 91t
FAST, *see* Friction assessment and screening test
Fatigue wear
    definition, 91t
    ferrogram particles, 24–25, 25f
    human teeth and, 154
    indications, 89t, 94t
    rolling contact fatigue, *see* Rolling contact
        fatigue
Ferrograms, 23–26, 24f, 25f
    abrasive wear particles, 23, 24f

adhesive wear particles, 23–24, 24f
contact fatigues wear particles, 23–24, 24f
normal particle from gearbox, 24, 25f
normal particle from turbine thrust, 24, 25f
particles from analytical, 25–26
Ferrographic analysis, 19t
Ferrography, 22–26
    areas of use, 24
    definition, 22
    developer, 22
    direct reading, 23
    ferrograms, 23–26, 24f, 25f
        abrasive wear particles, 23, 24f
        adhesive wear particles, 23–24, 24f
        contact fatigue wear particles, 23–24, 24f
        normal particle from gearbox, 24, 25f
        normal particle from turbine thrust, 24, 25f
        particles from analytical, 25–26
    schematic diagram, 23f
Ferrous particles, 25, 26
Fibers, 25
Fitch, Jim, 16
Flaking, 32, 33, 92t, 105f
Flank wear, 92t
Flash point test, 19t
Flow cavitation, 92t
Fluting, 32, 33, 92t
Four-ball lubricant tests, 165t, 166–167
Fourier transfer infrared spectroscopy (FTIR), 20t
Fourier transform, 15
Fracture, 14, 35, 53, 54f
Fretting wear (NR-Fr), 58–60
    ASM Committee classification, 32
    brinelling, 58
    cocoa, 90t
    common features among other types, 86t
    contact, 35, 58, 164t
    corrosion, 35, 92t
    definition, 58, 92t
    elastic contact radius, 59–60, 61f
    fatigue, 35
    Hertzian contact fatigue, 59–60, 60f
    kinds, 59, 59f
    MAC approach, 87
    nuclear power plant core components, 85
    occurrence in industry, 39t
    organizational diagram, 38f
    radial, 60f

tangential, 59, 59f
torsional, 59, 59f
wear, 35, 37, 87, 92t
wear testings, 164t
Frictional heating parameter, 147
Frictional wear, 38
Friction assessment and screening test (FAST), 168, 173
Friction coefficients, 143f, 168–171
Friction control, 8
Friction materials, 149
Friction problems, 9–10
Frosting, 67, 92t
FTIR, *see* Fourier transfer infrared spectroscopy
FTM 791 Method 3005.4, 26
Furrowing, 88f, 92t

G

Galling; *see also* Microgalling
    definition, 36–37, 36f, 92t
    illustrative examples, 52f, 119f
    quantification, 162
    single-occurrence mechanical damage, 34f, 35
    sliding contact and, 77f
    threshold stress, 162
    wear and, 8, 95t, 161
    wear testing, 164t
Gearboxes
    Dodge TDT 825 reduction, 24
    pulverizer, 23
    wear indications, 11t
    wind turbine
        chemical analysis, 17
        damages, 71f, 73f
        sketches, 104f, 105f
        TSA considerations, 153
Gear trains, 13
Gee, M. G., 45
German standard 50 322, 3–4
Gerotor cases, fluid pumps, 152
Gerotor pumps, 152
Glaeser, W. A., 18
Glazing, 93t
Goldstein, J. I., 123
Gouging abrasions, 34–35, 34f, 41, 43, 93t
Grand Canyon, 64, 65f, 66f
Graphite powder, 17
Greases
    cleaning, *see* Specimen cleaning
    definition, 17

deposits, 111f
particle analysis procedures, 26, 26t
screening, 166
silicon, 114
Gresham, R. M., 22
Grinding abrasions, 88f, 93t
Grinding action, 14
Grooves, 86t
Grooving 2-body abrasions (2BAb-Gr), 38f, 41, 88f, 93t, 105f
Gross reciprocating sliding wear (NR-GS), 38f, 57–58
Gross slip zones, 57, 58

H

H13 steel, 63f, 64f
Halmi, J., 71
Halo of damages, 86t
Hammering, 61, 94t
Hammers, 61, 62f
Hand lens, 106f
Hardware configuration and materials, 130t, 133
Hase, A., 14
HDL, *see* Hydrodynamic lubrication
HDR, *see* High Dynamic Range
Heat, 8, 9, 31
Heat checking, 33, 91t, 93t, 145
Helicopter blades, 66
Henderson, K. O., 16
Hertz equation, 59–60
Hertzian contact fatigue, 76–78
    definition, 93t
    fretting and, 59–60, 60f
    organizational diagram, 38f
    wear testing geometries, 165t
Hertz stress, 141
High Dynamic Range (HDR), 107
High-stress abrasions (2BAb-H)
    definition, 41, 43, 93t
    material dependence, 80
    nomenclature family, 88f
    organizational diagram, 38f
Hip joint, human, 61
Hitachi Instrument Company, 107
Hollow wear, 93t
Holm, R., 54
Human heart valve, 11t
Human hip joint, 61
Human teeth, 83, 153–154, 154f
Hybrid wear, 78–80

chemomechanical polishing wear (CMPol), 78; *see also* Chemomechanical polishing wear
organizational diagram, 38f
tribocorrosion (TrCor), 78–80; *see also* Tribocorrosion
Hydrodynamic lubrication (HDL), 142, 144

I

IBM Corporation, 62
Illumination, 114–116, 117t, 118t
Impact, 14
Impact abrasion, 94t
Impact wear, 62, 93t, 94t
Imperial College, 123
Impingement erosion, 94t
Inclined sphere-on-disk, 165t
Inconel alloy, 120t
Indentations, 32, 35, 86t
Indenting, 88f
Initial pitting, 33
Interface description, 130t, 134, 137
Interference wear, 32
Internal friction, 13
International Standards Organization (ISO), 149
IR laser scanning measuring system, 121t
ISO, *see* International Standards Organization
ISO 281:2007, 70
ISO 4406, 22
Ives, Lewis K., 162

J

Jackson, R. L., 142
Jaw crusher impact abrasion, 163t
Jaw crusher test, 43
Jet erosion, 94t; *see also* Open liquid jet erosion

K

Kaperick, J., 26
Kapoor, A., 57
Karl Fischer test, 20t
Kinematic wear marks, 94t

L

L_{10} lifetime, 70
Lacey, S. J., 13

Lambrechts, P., 79, 154
Lapping, 88f, 94t
Laser interferometry, 106f
Light optical microscopy (LOM), 106–116
 cross-sectioning, 112–113, 112f
 illumination, 114–116, 117t, 118t
 portable digital microscopes, 107, 108f, 109f
 range of magnifications, 106
 scanning acoustic microscopy (SAcM)
  versus, 125, 126f
 scanning electron microscopy (SEM) versus,
  103
 specimen cleaning, 110, 111f
 specimen mounting, 108–109
 taper-sectioning, 113–114, 114f, 115f
Lightweight rotors, 151–152
Linear oscillation, 59, 59f
Lingard, S., 14
Liquid droplet erosion (Er-LDr), 38f, 66, 164t
Liquid impingement erosion, 94t; *see also* Open
  liquid jet erosion
Load-bearing test system, 172f
Loading, 88f
Loads, 8, 140
Loop abrasion test, 163t
Loose components, 13
Low stress abrasions (2BAb-L), 38f, 41, 42f,
  88f, 94t
Lubricants; *see also* Lubrications
 chemical analysis, 17–18, 19t, 21t
 coin, 62
 liquid, 151
 loss of, 8
 reciprocating sliding wear, 55
 screening, 165t, 166–167
 solid, 17, 151
 TSA form, 132, 135, 141–142, 144
Lubrications; *see also* Lubricants
 condition, 130, 142
 depletion, 93t
 hydrodynamic, 142
 mixed film, 142
 Stribeck-based, 143
 TSA form, 130, 135
Ludema, Ken, 65, 174
Lundberg, Gustaf, 70

M

MAC, *see* Multiple attributes in context
McGhee-Tyson Airport, 50f
Machining defects, 13

Machining tools, 11t
McMahon, Matt, 25
Macropitting, 8, 67, 77f, 94t
Macrospalls, 72f, 81
Macrotribosystem, 2
Magic wear, 94t
Magnetic flux analysis, 20t
Map cracking, 91t
Martens, Adolf, 142
Martens-Stribeck curve, 142
Massachusetts, 120t
Matching, 166f
Materials
 dependence of wear patterns, 80–83
 loss, 36, 67, 96t
 surface methodology study, 120t
 transfer, 53, 86t, 89t
 TSA form, 130, 133, 137–138
Matted appearance, 86t
Maximum tribological scenario (XTS), 148, 149
May, Chris J., 16
MCSA, *see* Motor current signature analysis
MD, *see* Multidirectional sliding wear
Mechanical damage
 definition, 34
 diagram, 34f
 repetitive, 34f, 36–37
 single-occurrence, 34f, 35–36
Melting, 33, 34f
Meng, H. C., 65
Metal rollers, 12t
Methods
 surface measurement methodology study,
  121t
 wear problems detection
  acoustic emission (AE), 9–10, 9f, 11t,
   12t, 12–15
  motor current signature analysis
   (MCSA), 15
  oil analysis (OA), 15–26
  radionuclide wear detection, 27
  summary, 28t
  surface imaging, 10, 12
  vibration monitoring (VM), 9–10, 9f,
   11t, 12t, 12–15
  visual inspection, 10, 12
Michelin tires, 10, 12t
Microcracks, 77f, 86
Microgalling, 53
Micropitting
 contact fatigue and, 77f
 definition, 71, 94t

effects, 71–72
example, 73, 74f, 76, 77f
frosting and, 92t
quantification, 161–162
typical size, 73
wear testings, 165t
Microset™, 108
Microtribosystem, 2
Microwelding, 77f, 150
Mild wear, 95t
MIL-G-81322 method, 26
MIL-G-81937 method, 26
Mindlin, R. D., 59
Mindlin mechanical analysis, 58
Mining equipment, 4, 93t, 160
Miranda, João C., 84
Misalignment, 13
Mixed film lubrication, 142, 144
Mixed zones, 57, 58
Moderate wear, 32
Molybdenum disulfide, 17
Monochromatic illumination, 115–116
Motors, 11t
Motor current signature analysis (MCSA), 15
Multibody erosion, *see* Erosion, multibody
Multidirectional ball-in-socket, 164t
Multidirectional sliding wear (MD), 38f,
    60–61, 154, 164t
Multiple attributes in context (MAC), 85–87
Multiple light sources, 115

N

NCX 5102, 120t
Neanderthal man, 83
Nomarksi interference contrast, 118t
Nomenclature, *see* wear nomenclature and
    nomenclature families
Nomenclature families, definition, 88
Nominal tribological scenario (NTS), 148–149
Nonabrasions (S/N), 51–60, 95t
    examples, 52f, 53f, 54f
    organizational diagram, 38f
    overview, 51–54
    reciprocating sliding (NR), 55–60; *see also*
        Reciprocating sliding wear
        fretting (NR-FR), 58–60 (NR-FR);
            see also Fretting wear
        gross reciprocating sliding (NR-GS),
            57–58; *see also* Gross reciprocating
            sliding wear
        overview, 55–57

unidirectional sliding (NU), 54–55;
    see also Unidirectional sliding wear
Noncontact imaging and stylus profiling,
    see Stylus profiling and noncontact
    imaging
Nontraditional imaging methods, 124–127
    scanning acoustic microscopy (SAcM)
        components, 125f
        early uses, 125
        light optical microscopy (LOM) versus,
            125, 126f
        limitations, 127
        methodology, 124–125
    thermography and thermal wave imaging,
        127
Normal vibrational contact (VC), 76–78;
    see also Hertzian contact fatigue
Normal wear, 32, 84t, 95t
Norton-St. Gobain, 120t
Notching wear, 95t
NR-FR, *see* Fretting wear
NR-GS, *see* Gross reciprocating sliding
NU, *see* Unidirectional sliding wear
NTS, *see* Nominal tribological scenario

O

OA, *see* Oil analysis
Oak Ridge National Laboratory (ORNL), 44,
    123, 169, 169f, 172f
Oblique illumination, 117t
Off-brake wear, 149
Oil analysis (OA), 15–26
    chemical analysis of lubricants, 17–18, 19t,
        21t
    debris particle characterization, 18, 21f, 22
    definition, 15
    ferrography, 22–26
    filtration, 22–26
    magnetic traps, 22–26
    resources for analysis of wear problems, 99
    sampling, 16–17
        bad techniques, 16–17
        handling errors, 17
    significance, 16
    subject divisions, 15
    subject of focus, 16
    trending, 16
Oil detection paper, 20t
Open liquid jet erosion (Er/Jet-O), 38f, 68,
    164t
Open tribosystems, 4

Operating conditions, excess, 8
Operating environment, 130t, 132, 135
Optical diffusers, 115
Optical light pipe illuminator, 115
Optical microscopy, *see* Light optical microscopy
Ore-crushing equipment, 93t
Orientation contrast imaging (Orient-CI), 106f
ORNL, *see* Oak Ridge National Laboratory
Oxidative wear, 95t

P

Palmgren, Arvid, 70
Particle counting, 20t
Peening wear, 95t
Peterson, M. B., 87, 162
Phase contrast illumination, 118t
Pin abrasion test, 163t
Pin-on-disk sliding wear, 54f, 163t
Piston ring groove wear, 150
Pits, fine-scale, 86t
Pivoting motion, 59, 59f
Plastic bearings, 146
Plastic deformation
    consistent terminology, 38
    galling and, 92t
    illustrative examples, 53, 53f, 70, 71f, 154, 154f
    scuffing and, 96t
    severe (metallic) wear, 96t
    sliding contact and, 77f, 105f
    smearing and, 97t
Plastic smearing, 95t
Plowing (ploughing), 88, 88f, 95t
Polarized light (PL)
    illumination, 117t
    microscopy, 106f
Polishing wear (S/Ab/Pol)
    chemochemical (CMPol), 49–51, 78
    definition, 95t
    nomenclature family, 88f
    organizational diagram, 38f
    sliding contact and, 77f
Polychromatic illumination, 115–116
Polycrystalline ceramic tile, 120t
Polytetrafluoroethylene (PTFE), 17, 114
Portable digital microscopes, 107, 108f, 109f
Problem description, 130t, 132, 136
Progressive wear, 50, 84t
PTFE, *see* Polytetrafluoroethylene
Pure rolling contact fatigue (RCF-P), 69–73

achieving, 69
organizational diagram, 38f
rolling element bearings, 69–70, 71f
wear testings, 165t
PV limit, 149

Q

Quench cracking, 33

R

Radial expansion and contraction, 59, 59f, 60f
Radionuclide wear detection, 27
Radius, elastic contact, 61f
Railroad wheels, 93t
Rain erosion, 66, 95t
Rake face wear, 95t
Ramalho, Amilcar, 84
Raymond 783 RPS, 23
RCF-P, *see* Pure rolling contact fatigue
RCF-Slp, *see* Rolling contact fatigue with slip
Reciprocating pin-on-flat, 163t
Reciprocating sliding wear (NR), 55–60
    fretting (NR-FR), 58–60; *see also* Fretting wear
    gross reciprocating sliding (NR-GS), 57–58; *see also* Gross reciprocating sliding wear
    organizational diagram, 38f
    overview, 55–57
    wear testings, 163t, 164t
Red mud, 60, 87, 96t
Red tint plate, 117t
Reduced life, 84t
Relative motions, 34f, 38f
Rennselar, J. Van, 70
Repetitive impact (RI), 61–69
    multibody erosion (Er), 64–69; *see also* Erosion, multibody
    cavitation erosion, 67, 68f
    erodants, 64
    erosion factors, 65
    extensity of erosion, 64
    liquid droplet erosion, 66
    open liquid jet erosion, 68
    slurry erosion, 66–67
    solid impingement erosion, 66
    spark erosion, 67
    steam erosion, 69, 69f
    submerged jet erosion, 68

organizational diagram, 38f
2-body repetitive impact (RI/2B or RI/3B),
  61–64; *see also* 2-body repetitive
  impact
  definition, 61
  effects, cost, and economics, 62
  examples, 61–62, 62f
  human teeth, 154
  organizational diagram, 38f
  properties of materials, 62
  3-body abrasions and, 62–64, 63f, 64f
Resonance, 13
RI/2B, *see* Repetitive impact
RI/3B, *see* Repetitive impact
Ridging, 96t
Ring-on-ring, 164t
Rodenstock, 121t
Roller deterioration, 14t
Roll formation, 96t
Rolling contact fatigue (RCF)
  definition, 96t
  organizational diagram, 38f
  pure (RCF-P), 69–73; *see also* Pure rolling
    contact fatigue
  with slip (RCF-Slp), 73–76; *see also* Rolling
    contact fatigue with slip
  wear testing, 165t
Rolling contact fatigue with slip (RCF-Slp)
  common features among other types, 86t
  hairline tensile cracks, 73, 75f, 75
  microcracks, 75, 76f
  organizational diagram, 38f
  spalling, 73, 74f
  wear testings, 164t, 165t
  white-etching layers, 75, 76f
Rolling element bearings
  contact fatigue and, 69–70, 71f
  significance, 13, 15
  spark erosions, 67
  wear indications, 11t
  wear types, 32, 58
Rotational fretting, 59, 59f
Rubbing action, 13
Running-in, 84t
Ryason, P. R., 49

S

S (Sliding contact), *see* Sliding contact
S/2-3Ab, *see* Combined abrasions
S/Ab, *see* Abrasions
S/Ab/Pol, *see* Polishing wear

S/N, *see* Nonabrasions
SAD (Selected area diffraction), 106f
SAE, *see* Society of Automotive Engineers
SAM (Scanning acoustic microscopy), 106f
Samuels, L. E., 49, 113
Sanding wear, 61, 88f
Sandpaper, 41
Sand
  silica, 41, 42, 47
  Grand canyon, 34, 35f
  normal illumination vs. HDR, 109f
  wear testing, 163t
Scanning acoustic microscopy (SAcM)
  components, 125f
  early uses, 125
  hierarchy diagram, 106f
  light optical microscopy (LOM) versus, 125,
    126f
  limitations, 127
  methodology, 124–125
Scanning electron microscopy (SEM), 18, 47f,
  115
Scanning electron microscopy (SEM), 103,
  123–124
Scanning transmission electron microscopy
  (STEM), 106f
Scoring
  ASM classifications, 32
  consistent terminology, 38
  definition, 36, 96t
  nomenclature, 88f
  rolling contact damage, 73f
  sliding contact, 77f
  wear indications, 9f, 11t
Scratches
  detecting, 117t
  illustrative examples, 42f, 52f, 53, 53f,
    154f
  kinematic wear marks, 94t
  multidirectional, 86t
  polishing wear, 49, 50
  single mechanical event, 34f, 35
  single-point scratch test, 43–45
    Al_2O_3 and SiC, single phases, 45f
    Al_2O_3-SiC composite, 46f
    pure Fe, Ni, and Co, 43f
    types and severity of damage, 44t
  3-body abrasions, 49
  wear indications, 9f
Scuffing wear
  ASM classification, 32
  consistent terminology, 38

definition, 38, 96t
fuel injector plungers, 153, 154
gerotor pumps, 152
illustrative examples, 50f, 52–53, 52f, 71f
nomenclature family, 88f
nonabrasive wear, 95t
pistons, 58
sliding contact and, 77f
smoothing process, 116
TSA form, 152
wear indications, 8f
Seizure, 35, 77f, 96t, 97t, 136
Selected area diffraction (SAD), 106f
SEM, *see* Scanning electron microscope
Sensitive tint illumination, 117t
Severe (metallic) wear, 96t
Shear, ductile, 86t
Shelling, 96t
SiC abrasive paper, 47f, 48f
Silica sand, 41, 42f
Silicon nitride, 53, 120t, 122f
Silicon wafer manufacturing, 50
Silver, 17
Singh, Sukhjeet, 15
Single-point scratch test, 43–45
    Al_2O_3 and SiC, single phases, 45f
    Al_2O_3-SiC composite, 46f
    pure Fe, Ni, and Co, 43f
    types and severity of damage, 44t
Sipes, 80, 81f
Skidding, 32
Skirt areas, 58
SLA, *see* Surface layer activation
Sliding contact (S), 41–61
    abrasive wear, 41–51
        abradants, 41, 42f
        ASM classification, 32, 33
        definition, ASTM G40-13b, 41
        ferrogram particles, 23, 24f
        occurrence in industry, 39t
        organizational diagram, 38f
        3-body abrasive wear, 47, 49–51
        2-body abrasive wear, 41–47
    definition, 41
    multidirectional sliding, 60–61
    nonabrasive wear, 51–60
        examples, 52f, 53f, 54f
        gross reciprocating sliding, 57–58
        organizational diagram, 38f
        overview, 51–54
        reciprocating sliding, 55–60
        unidirectional sliding, 54–55

organizational diagram, 34f, 38f
    wear regimes, 84t
Sliding face seal, 163t
Sliding wear; *see also* Sliding contact
    block-on-ring, 163t
    definition, 97t
    pin-on-disk, 54f, 163t
Slurry, definition, 67
Slurry abrasivity, 163t
Slurry erosion (Er-Slu), 38f, 66–67, 97t
Smearing
    ASM classification, 32
    consistent terminology, 38
    definition, 97t
    illustrative examples, 104f
    nonabrasive wear, 95t
    pure rolling contact fatigue, 69–70
SOAP, *see* Spectroscopic Oil Analysis Program
Society of Automotive Engineers (SAE), 149
Softening, 32, 33
Solar panels, 11t
Solid impingement erosion (Er-Slm), 38f, 66,
    86t, 164t
Solid lubricants, 17, 151
Song, J. F., 116
Spalling
    advanced form, 96t
    ASM wear classification, 32, 33
    causes of wear, 8
    contact fatigue and, 77f
    definition, 37, 38, 97t
    illustrative examples, 70, 71f, 73, 74f, 104f,
        105f
    macroscale, 86t
    micropitting and, 72–73
    wear indications, 9f
Spark erosion (Er-Spk), 38f, 67, 97t, 164t
Specimen cleaning, 110, 111f
Specimen mounting, 108–109
Spectrometric analysis, 20t
Spectroscopic Oil Analysis Program (SOAP),
    17
Speed, excessive, 8
Spurr, R. T., 47
SSBT, *see* Subscale brake materials testing
Stamping, 61
Steam erosion (Er-Stm), 38f, 69, 69f
Steel
    AISI 304 stainless steel, 50
    AISI 440XH stainless steel, 73
    AISI 8620 steel, 171, 172t
    AISI 52100 steel, 47f, 48f, 70

H13 steel, 63f, 64f
Type 440HX stainless steel, 76f
STEM (scanning transmission electron
microscopy), 106f
Stereo macroscope, 31
Stick slip, 97t
Stick zones, 57, 58
Striations, 41, 86t
Stribeck, Richard, 142
Stribeck curve, 142–144, 143f
Stylus profiling and noncontact imaging, 116,
118–123
benefits and limitations, 118–120
confocal microscopy, 121, 123
uses, 116
Sublimation, 33
Submarine shaft seals, 11t
Submerged liquid jet erosion (Er/Jet-S), 38f,
68, 164t
Subscale brake materials testing (SSBT), 169,
169f, 170, 171f
Suh, Nam P., 55
Surface appearance, 9, 9f
Surface damages and wear, 31–99
consistent terminology, 32–33
surface damage types, 33–37
chemical reaction, 33, 34f
dissolution, 33, 34f
examples of concurrent, 34–35
mechanical damage, 34f, 34
organizational diagram, 34f
radiation damage, 33, 34f
thermal damage, 33, 34f
wear diagnosis, 85–87
wear nomenclature and nomenclature
families, 87–98, 98t
wear patterns and surface appearance,
80–83
wear problems, resources for analysis, 98–99
wear transitions, 83–85
wear types and characteristics, 37–80
ASM Committee, 32–33
categories, processes, and mechanisms,
40, 40f
overview of categorization, 37–39
relative motion system, 40–80
Surface dissolution, 33
Surface distress, 97t
Surface fatigue, 32, 33
Surface layer activation (SLA), 27
Surface layers, delamination, 86t
Surface roughness, 120–121, 120t, 121t, 122f

T

2BAb, *see* 2-body abrasions
2BAb-Gr, *see* Grooving abrasions
2BAb-H, *see* High-stress abrasions
2BAb-L, *see* Low-stress abrasions
3BAb, *see* 3-body abrasions
Taber Abraser, 173–174
Taber index (TI), 163t, 174
Tabor, David, 52, 139
Talysurf 10™, 121t
Tangential fretting, 59, 59f
Taper-sectioning, 113–114, 114f, 115f
Taylor, C. M., 57
Tearing, 43, 53, 86t
Technology and Maintenance Council (TMC),
149
Teflon®, 17, 113
Temperature, 141
Tennessee Highway Patrol, 169
Tensile cracking, 97t
Thermal cycling, 91t
Thermography, 127
3-body abrasions (3BAb), 49–51
abradants, 42f
combined 2-body abrasive wear (S/2-3Ab),
51, 51f
common features among other types, 86t
definition, 97t
human teeth and, 154
nomenclature families, 88t
organizational diagram, 38f
polishing wear, 38f, 49–51
removal of material, 94t
sliding contact and, 77f
3-body abrasion test, 47
2-body repetitive impact and, 62–64, 63f,
64f
wear testings, 163t
Threshold stress, 162
TI, *see* Taber index
Ti-6Al-V4 alloy, 42f, 171, 172t
Tires; *see also* Brakes
Michelin tires, 10, 12t
wear indications, 10, 12t, 81, 82t
wear patterns, 80, 81f, 81, 82t
Titanium alloys, 42f, 151–152
TMC, *see* Technology and Maintenance
Council
Tools, wear surfaces imaging and
characterization, 103–127
cross-sectional viewing, 106

human eye, 104
light and electron optical imaging tools,
106f
light optical microscopy (LOM), 106–116
cross-sectioning, 112–113, 112f
illumination, 114–116, 117t, 118t
portable digital microscopes, 107, 108f,
109f
range of magnifications, 106
scanning electron microscopy (SEM)
versus, 103
specimen cleaning, 110, 111f
specimen mounting, 108–109
taper-sectioning, 113–114, 114f, 115f
most commonly used, 103
nontraditional imaging methods, 124–127
scanning acoustic microscopy (SAcM),
124–125, 125f, 126f, 127
thermography and thermal wave
imaging, 127
normal surface viewing, 106
scanning electron microscopy (SEM),
123–124
sketching, 104, 104f, 105f
stylus profiling and noncontact imaging,
116, 118–123
benefits and limitations, 118–120
confocal microscopy, 121, 123
surface roughness, 120–121, 122f
uses, 116
transmission electron microscopy (TEM),
106f, 123–124
visual examination, 103
Topometrix, 121t
Torching phenomenon, 151
Torsional fretting, 59, 59f
Total acid number, 20t
Total base number, 21t
Transducers, 13
Transfer, 97t
adhesive, 52f, 77f
Transmission electron microscopy (TEM),
106f, 123–124
Transmitted light illumination, 114
Transportation Research Center (TRC), 169,
170f
TRC, *see* Transportation Research Center
Tread block, 80, 81f
Tread depth, 10, 12t
Trending, 16
Tribochemical wear, 79, 154
Tribocorrosion (TrCor), 78–80

definition, 34, 78, 98t
dental erosion, 79
MAC approach, 87
organizational diagram, 38f
sequence, 79t
slurry and, 67
Triboelements, 1–2, 137
Tribology, xi, 1, 31
Tribosystem analysis (TSA), xii, xiii, 7;
see also Tribosystem analysis form;
Tribosystem analysis matching
Tribosystem analysis (TSA) form, 129–154
explanation of entries, 132–148
description of the components,
geometry, and materials, 137–138
description of the operating conditions,
138–144
illustrative examples of form parts,
132–136
problem description, 144–148
missing information, 148–150
overview, 130–132
section parts
Section heading, 130t, 133
Section 1 (Hardware configuration and
materials), 130t, 133
Section 1.1 (Interface description), 130t,
134, 137
Block 1.1, 137–138
Block 1.2, 137–138
Block 1.3, 137–138
Block 1.4, 137–138
Block 1.5, 138
section 2 (Operating environment),
130t, 132, 135
Block 2.1, 132, 135, 138
Block 2.2, 132, 135, 138–139
Block 2.3, 132, 135, 139–141
Block 2.4, 132, 135, 141
Block 2.5, 132, 135, 141–142
Block 2.6, 132, 135, 141–142
Block 2.7, 132, 135, 142–144
Block 2.8, 132, 135, 144
Block 2.9, 132, 135, 144
Section 3 (Problem description), 130t,
132, 136
Block 3.1, 145
Block 3.2, 132, 145
Block 3.3, 132, 145–147
Block 3.4, 132, 147
Block 3.5, 132, 147
Block 3.6, 148

Section 4 (Attachment and exhibits),
130t, 131, 136
summary of sections, 130t
tailoring to specific problems, 150–154
diesel engine valve/valve seat wear, 151
fuel injector plungers for low-sulfur
fuels, 152–153
gerotor cases for fluid pumps, 152
human teeth, 153–154, 154f
lightweight rotors for truck brakes,
151–152
piston ring groove wear, 150
vanes for diesel engine turbochargers,
150–151
wind turbine gearbox bearings, 153
Tribosystem analysis (TSA) matching,
165–173
accelerated testing, 167–168
Case Study 1: friction coefficient
versus field performance for brake
lining materials, 168–171
Case Study 2: spectrum loading
tribotests, 171–172
schematic method, 165–167, 166f
variable-condition versus constant-
condition tribotests, 173
Tribosystems
closed, 4
complex, 2
diagnosing lubricant problems, 4
definition, 1
establishing boundaries, 2
example, 2
levels, 3, 3f
macrotribosystem, 2
microtribosystem, 2
open, 4
triboelements, 1–2
wear problem examples, 11t, 12t
Tribotestings; *see also* Tribosystem analysis
matching
common wear testing geometries, 163t,
165t
overview, 160
purposes (levels), 4, 161–162
significance, 162
TrCor, *see* Tribocorrosion
Troyer, Drew, 16
True brinelling, 32, 58
TSA, *see* Tribosystem Analysis
Tumbling, 88f

Twisting contact, 59f
2-body abrasions (2BAb), 41–47
abradants, 41, 43
combined 3-body abrasive wear (S/2-3Ab),
51, 51f
common features among other types, 86t
debris, 45–46, 48f
definition, 41, 98t
full contact, 44–45, 47f
gouging abrasions, 41, 43
grooves, 41, 43f, 47f
grooving abrasions, 41
high-stress abrasions, 41
human teeth and, 154
jaw crusher test, 43
low stress abrasions, 41, 42f, 46–47
nomenclature families, 88t
organizational diagram, 38f
proposed mechanism, 41
sandpaper, 41
single-point scratch test, 43–45
Al_2O_3 and SiC, single phases, 45f
Al_2O_3-SiC composite, 46f
pure Fe, Ni, and Co, 43f
types and severity of damage, 44t
sliding contact and, 77
striations, 41
wear testings, 163t
2-body repetitive impact (RI/2B or RI/3B),
61–64
definition, 61
effects, cost, and economics, 62
examples, 61–62, 62f
human teeth, 154
organizational diagram, 38f
properties of materials, 62
3-body abrasions and, 62–64, 63f, 64f
wear testings, 163t
Type 440HX stainless steel, 76f

U

Unbalance, 13
Unidirectional sliding wear (NU), 54–55
common features among other types, 86t
organizational diagram, 38f
wear testings, 163t, 164t, 165t
U.S. Department of Transportation, 169
U.S. Federal Motor Vehicle Safety Standard
FMVSS-121, 169
U.S. Space Shuttle *Discovery* mission, 80, 81f

V

Vane pivots, 149
Vanes, diesel engine turbochargers, 150–151
VC, *see* Normal vibrational contact
Vertical scanning interferometry, 118t
Vibration monitoring (VM), 9–10, 9f, 11t, 12t, 12–15
Vibrations, 8, 13, 61
Vibratory cavitation, 98t
Viscosity, 21t, 56, 85, 135, 142, 143
Visual inspection, 10, 12, 103
VM, *see* Vibration monitoring
Voort, Vander, 116
Vorburger, T. V., 116

W

Waterhouse, R. B., 58
Wavelet transform, 15
Wear
    by-products, 9, 9f
    categories, 40, 40f
    codes for categories and subcategories, 38f
    common features among types, 86t
    debris, 18, 26
    definition, 34f, 36, 98t, 116
    diagnosis, 31, 85–87
    factors, 84
    hybrid forms, 38f
    mechanisms, 40, 40f
    pattern charts, 10
    problems, *see* Wear problems
    problem solving, *see* Wear problem solving
    processes, 40, 40f
    types
        ASM Committee, 32–33
        common features among, 86t
        consistent terminology, 32–33
        occurrence in industry, 39t
        relative motion, *see* Wear types (relative motion)
Wear Check Canada International, 23
Wear diagnosis, 31, 85–87
Wear-in, 84t
Wear nomenclature and nomenclature families, 87–98, 98t
    nomenclature used in wear literature, 88t, 98t
    overview, 87–88
Wear-out, 84t
Wear pattern charts, 10

Wear patterns
    charts, 10
    human teeth, 83
    significance, 80
    surface appearance and, 80–83
    tire, 80, 81f, 81, 82t
Wear problems, 7–29
    causes, 8
    control, 8
    current approach of diagnosis, 31
    direct indications, 9, 9f, 10, 11t, 12t
    examples, 10, 11t, 12t
    methods of detection, 10–27
        acoustic emission (AE), 9–10, 9f, 11t, 12t, 12–15
        motor current signature analysis (MCSA), 15
        oil analysis (OA), 15–26
        radionuclide wear detection, 27
        summary, 28t
        surface imaging, 10, 12
        vibration monitoring (VM), 9–10, 9f, 11t, 12t, 12–15
        visual inspection, 10, 12
    presentations, 9–10, 9f
    proper function of materials, 8
    resources for analysis, 98–99
        images of wear surfaces and failed components, 98–99
        oil analysis and functions of additives, 99
        wear debris appearance, 99
    superficial indications, 7
    surface damages and, 31
Wear problem solving, 157–175
    options for, 158–160
    overview, 157–158
    tribotestings, 160–165
        common wear testing geometries, 163t, 165t
        overview, 160
        purposes, 161–162
        significance, 162
    TSA matching, 165–173
        accelerated testing, 167–168
        Case Study 1: friction coefficient versus field performance for brake lining materials, 168–171
        Case Study 2: spectrum loading tribotests, 171–172
        schematic method, 165–167, 166f

variable-condition versus constant-condition tribotests, 173
wear testing, 173–175
Wear surfaces
  resources for analysis of wear problems, 98–99
  tools for imaging and characterization, 103–127
    light optical microscopy (LOM), 106–116
    nontraditional imaging methods, 124–127
    overview, 103–106
    scanning electron microscopy (SEM), 123–124
    stylus profiling and noncontact imaging, 116, 119–123
    transmission electron microscopy (TEM), 123–124
Wear testings, purposes, 161
Wear transitions, 83–85, 84t
Wear types; *see also specific wear types*
  ASM Committee, 32–33
  common features among, 86t
  consistent terminology, 32–33
  contact fatigue (CF), 69–78
    normal vibrational contact (VC), 76–78
    pure rolling contact fatigue (RCF-P), 69–73
    rolling contact fatigue with slip (RCF-Slp)
  hybrid, 78–80
  repetitive impact (RI), 61–69
    multibody erosion, 64–69

    2-body repetitive impact, 61–64
  sliding contact (S), 41–61
    abrasive wear, 41–51
    definition, 41
    multidirectional sliding, 60–61
    nonabrasive wear, 51–60
Weathering, 65f
Wescott, V. C., 22
Wet sand/rubber wheel test, 163t
Wet sand/steel wheel test, 163t
White-etching layers, 75, 76f
White nonferrous particles, 25
Whitenton, Eric P., 162
White reflectors, 115
Wind turbine gearboxes
  chemical analysis, 17
  damages, 71f, 73f
  sketches, 104f, 105f
  TSA considerations, 153
Winer, W. O., 87
Worcester, 120t
Work rate, 147

X

Xin, J., 51
XTS, *see* Maximum tribological scenario

Z

Zaretsky, 70
Zhou, Z. R., 58, 59
Zhu, M. H., 59

Milton Keynes UK
Ingram Content Group UK Ltd.
UKHW040057071024
449327UK00019B/627